Problems Supplement to accompany
Vector Mechanics for Engineers: Dynamics
Sixth Edition

Problems Supplement

to accompany

Vector Mechanics for Engineers: Dynamics

Sixth Edition

Edited by

William E. Clausen
The Ohio State University

Amos Gilat
The Ohio State University

Boston Burr Ridge, IL Dubuque, IA Madison, WI New York San Francisco St. Louis
Bangkok Bogotá Caracas Kuala Lumpur Lisbon London Madrid Mexico City
Milan Montreal New Delhi Santiago Seoul Singapore Sydney Taipei Toronto

McGraw-Hill Higher Education

A Division of The McGraw-Hill Companies

Problems Supplement to accompany
VECTOR MECHANICS FOR ENGINEERS: DYNAMICS
Beer and Johnston

Published by McGraw-Hill, an imprint of the McGraw-Hill Companies, Inc., 1221 Avenue of the Americas, New York, NY 10020.

1 2 3 4 5 6 7 8 9 0 QPD/QPD 0 9 8 7 6 5 4 3 2 1 0

ISBN 0-07-244351-0

www.mhhe.com

TABLE OF CONTENTS

CHAPTER 11
KINEMATICS OF PARTICLES

SECTIONS 11.1 to 11.3

11.1 The motion of a particle is defined by the relation $x = 2t^3 - 15t^2 + 36t - 10$, where x is expressed in meters and t in seconds. Determine the position, velocity, and acceleration when $t = 4$ s.

11.2 The motion of a particle is defined by the relation $x = t^3 - 3t^2 + 6$, where x is expressed in feet and t in seconds. Determine the time, position, and acceleration when $v = 0$.

11.3 The motion of a particle is defined by the relation $x = t^3 - 9t^2 + 15t + 18$, where x is expressed in meters and t in seconds. Determine the time, position, and acceleration when $v = 0$.

11.4 The acceleration of a particle is defined by the relation $a = -5\ \text{ft/s}^2$. If $v = +30$ ft/s and $x = 0$ when $t = 0$, determine the velocity, position, and total distance traveled when $t = 8$ s.

11.5 The acceleration of a particle is defined by the relation $a = 32 - 6t^2$. The particle starts at $t = 0$ with $v = 0$ and $x = 50$ m. Determine (*a*) the time when the velocity is again zero, (*b*) the position and velocity when $t = 6$ s, (*c*) the total distance traveled by the particle from $t = 0$ to $t = 6$ s.

11.6 The acceleration of a particle is defined by the relation $a = kt^2$. (*a*) Knowing that $v = -24$ ft/s when $t = 0$ and that $v = +40$ ft/s when $t = 4$ s, determine the constant k. (*b*) Write the equations of motion, knowing also that $x = 6$ ft when $t = 2$ s.

11.7 The acceleration of a particle is defined by the relation $a = -kx^{-2}$. The particle starts with no initial velocity at $x = 900$ mm, and it is observed that its velocity is 10 m/s when $x = 300$ mm. Determine (*a*) the value of k, (*b*) the velocity of the particle when $x = 500$ mm.

11.8 The acceleration of a particle is defined by the relation $a = -k/x$. It has been experimentally determined that $v = 4$ m/s when $x = 250$ mm and that $v = 3$ m/s when $x = 500$ mm. Determine (*a*) the velocity of the particle when $x = 750$ mm, (*b*) the position of the particle at which its velocity is zero.

11.9 The acceleration of an oscillating particle is defined by the relation $a = -kx$. Find the value of k such that $v = 24$ in./s when $x = 0$ and $x = 6$ in. when $v = 0$.

11.10 The acceleration of a particle is defined by the relation $a = 90 - 6x^2$, where a is expressed in in./s^2 and x in inches. The particle starts with no initial velocity at the position $x = 0$. Determine (a) the velocity when $x = 5$ in., (b) the position where the velocity is again zero, (c) the position where the velocity is maximum.

11.11 The acceleration of a particle is defined by the relation $a = -3v$, where a is expressed in in./s^2 and v in in./s. Knowing that at $t = 0$ the velocity is 60 in./s, determine (a) the distance the particle will travel before coming to rest, (b) the time required for the particle to come to rest, (c) the time required for the velocity of the particle to be reduced to 1 percent of its initial value.

11.12 The acceleration of a particle is defined by the relation $a = -kv^2$, where a is expressed in ft/s^2 and v in ft/s. The particle starts at $x = 0$ with a velocity of 20 ft/s and when $x = 100$ ft the velocity is found to be 15 ft/s. Determine the distance the particle will travel (a) before its velocity drops to 10 ft/s, (b) before it comes to rest.

11.13 The acceleration of a particle is defined by the relation $a = -kv^{1.5}$. The particle starts at $t = 0$ and $x = 0$ with an initial velocity v_0. (a) Show that the velocity and position coordinate at any time t are related by the equation $x/t = \sqrt{v_0 v}$. (b) Knowing that for $v_0 = 36$ m/s the particle comes to rest after traveling 3 m, determine the velocity of the particle and the time when $x = 2$ m.

11.14 The velocity of a particle is defined by the relation $v = 40 - 0.2x$, where v is expressed in m/s and x in meters. Knowing that $x = 0$ at $t = 0$, determine (a) the distance traveled before the particle comes to rest, (b) the acceleration at $t = 0$, (c) the time when $x = 50$ m.

11.15 A projectile enters a resisting medium at $x = 0$ with an initial velocity $v_0 = 900$ ft/s and travels 4 in. before coming to rest. Assuming that the velocity of the projectile was defined by the relation $v = v_0 - kx$, where v is expressed in ft/s and x in feet, determine (a) the initial acceleration of the projectile, (b) the time required for the projectile to penetrate 3.9 in. into the resisting medium.

11.16 The acceleration of a particle is $a = k \sin(\pi t/T)$. Knowing that both the velocity and the position coordinate of the particle are zero when $t = 0$, determine (a) the equations of motion, (b) the maximum velocity, (c) the position at $t = 2T$, (d) the average velocity during the interval $t = 0$ to $t = 2T$.

SECTIONS 11.4 to 11.6

11.17 A ball is thrown vertically upward from a point on a tower located 25 m above the ground. Knowing that the ball strikes the ground 3 s after release, determine the speed with which the ball (a) was thrown upward, (b) strikes the ground.

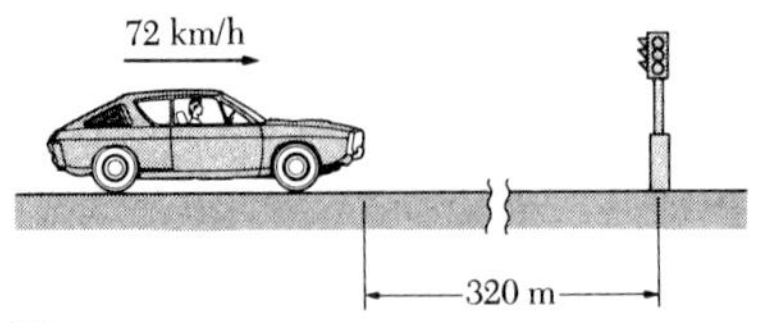

Fig. P11.18

11.18 A motorist is traveling at 72 km/h when she observes that a traffic light 320 m ahead of her turns red. The traffic light is timed to stay red for 22 s. If the motorist wishes to pass the light without stopping just as it turns green again, determine (a) the required uniform deceleration of the car, (b) the speed of the car as it passes the light.

11.19 An automobile travels 800 ft in 20 s while being accelerated at a constant rate of 2.5 ft/s². Determine (*a*) its initial velocity, (*b*) its final velocity, (*c*) the distance traveled during the first 10 s.

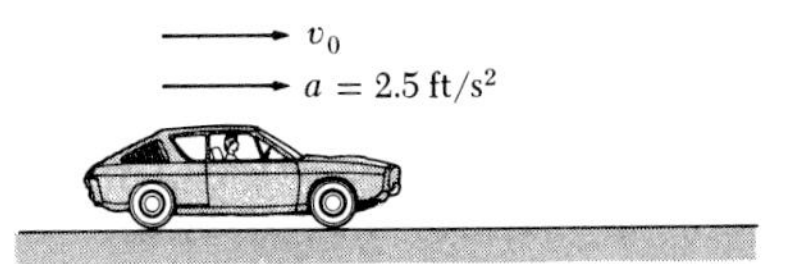

Fig. P11.19

11. 20 A stone is released from an elevator moving up at a speed of 12 ft/s and reaches the bottom of the shaft in 2.5 s. (*a*) How high was the elevator when the stone was released? (*b*) With what speed does the stone strike the bottom of the shaft?

11.21 A bus is accelerated at the rate of 1.2 m/s² as it travels from *A* to *B*. Knowing that the speed of the bus was $v_0 = 18$ km/h as it passed *A*, determine (*a*) the time required for the bus to reach *B*, (*b*) the corresponding speed as it passes *B*.

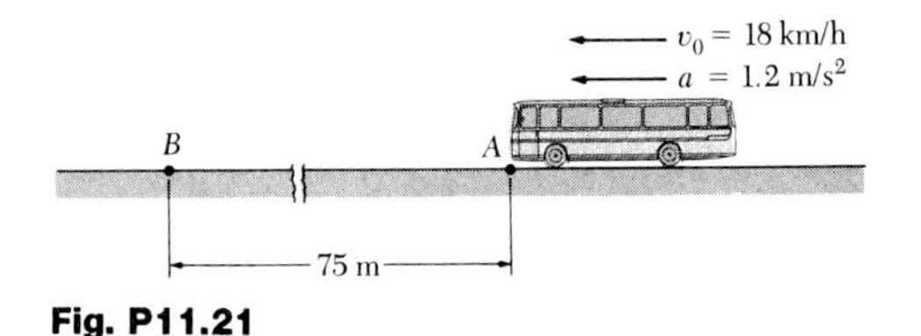

Fig. P11.21

11.22 Automobile *A* starts from *O* and accelerates at the constant rate of 0.8 m/s². A short time later it is passed by bus *B* which is traveling in the opposite direction at a constant speed of 5 m/s. Knowing that bus *B* passes point *O* 22 s after automobile *A* started from there, determine when and where the vehicles passed each other.

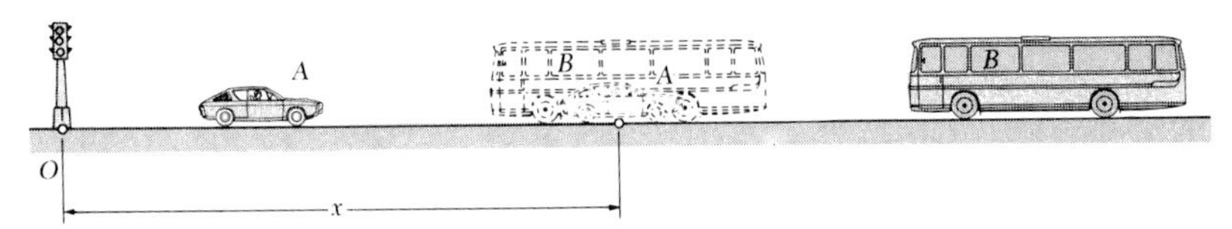

Fig. P11.22

11.23 An open-platform elevator is moving down a mine shaft at a constant velocity v_e when the elevator platform hits and dislodges a stone. (*a*) Assuming that the stone starts falling with no initial velocity, show that the stone will hit the platform with a relative velocity of magnitude v_e. (*b*) If $v_e = 7.5$ m/s, determine when and where the stone will hit the elevator platform.

11.24 A freight elevator moving upward with a constant velocity of 6 ft/s passes a passenger elevator which is stopped. Four seconds later the passenger elevator starts upward with a constant acceleration of 2.4 ft/s². Determine (*a*) when and where the elevators will be at the same height, (*b*) the speed of the passenger elevator at that time.

11.25 Automobiles *A* and *B* are traveling in adjacent highway lanes and at $t = 0$ have the positions and speeds shown. Knowing that automobile *A* has a constant acceleration of 2 ft/s² and that *B* has a constant deceleration of 1.5 ft/s², determine (*a*) when and where *A* will overtake *B*, (*b*) the speed of each automobile at that time.

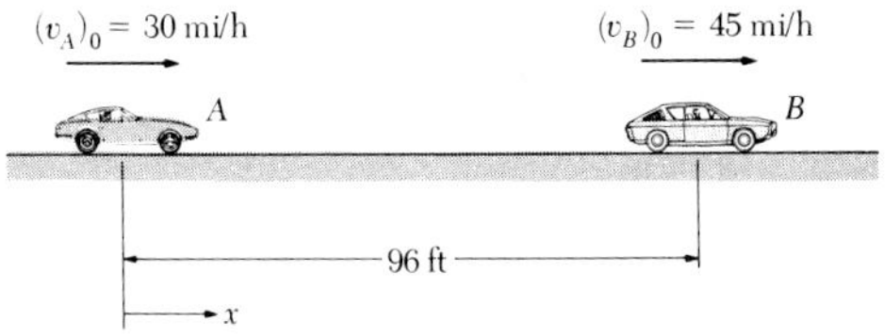

$(v_A)_0 = 30$ mi/h $(v_B)_0 = 45$ mi/h

A B

96 ft

x

Fig. P11.25

11.26 The elevator shown in the figure moves upward at the constant velocity of 4 m/s. Determine (*a*) the velocity of the cable *C*, (*b*) the velocity of the counterweight *W*, (*c*) the relative velocity of the cable *C* with respect to the elevator, (*d*) the relative velocity of the counterweight *W* with respect to the elevator.

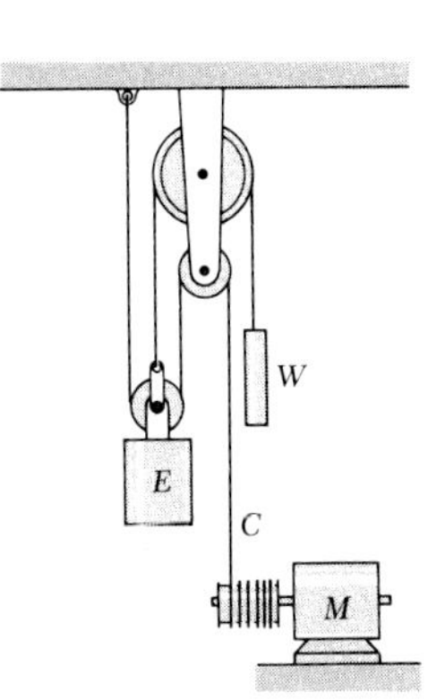

Fig. P11.26

11.27 The slider block *B* moves to the right with a constant velocity of 20 in./s. Determine (*a*) the velocity of block *A*, (*b*) the velocity of portion *D* of the cable, (*c*) the relative velocity of *A* with respect to *B*, (*d*) the relative velocity of portion *C* of the cable with respect to portion *D*.

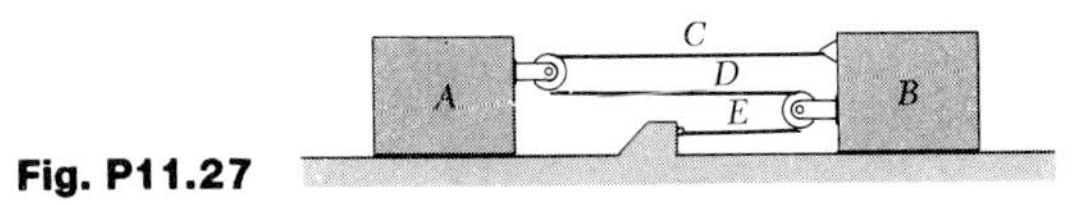

Fig. P11.27

11.28 The slider block A starts from rest and moves to the left with a constant acceleration. Knowing that the velocity of block B is 12 in./s after moving 24 in., determine (a) the accelerations of A and B, (b) the velocity and position of A after 5 s.

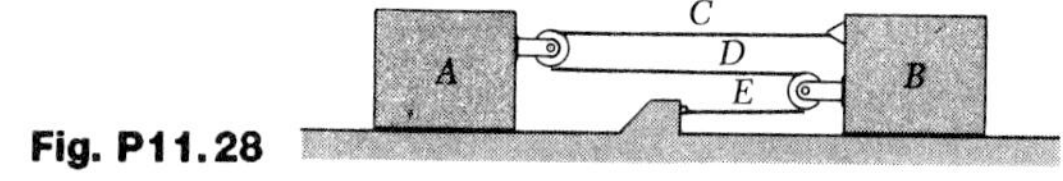

Fig. P11.28

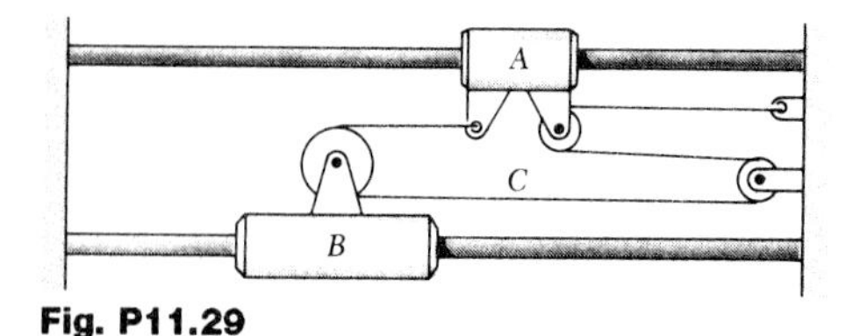

Fig. P11.29

11.29 Collar A starts from rest and moves to the left with a constant acceleration. Knowing that after 4 s the relative velocity of collar B with respect to collar A is 300 mm/s, determine (a) the accelerations of A and B, (b) the position and velocity of B after 5 s.

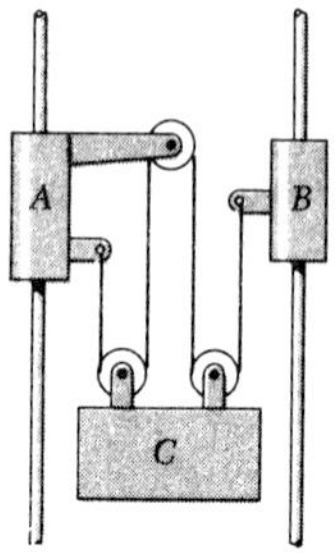

Fig. P11.30 and P11.31

11.30 Collar A starts from rest at $t = 0$ and moves upward with a constant acceleration of 2.5 in./s^2. Knowing that collar B moves downward with a constant velocity of 15 in./s, determine (a) the time at which the velocity of block C is zero, (b) the corresponding position of block C.

11.31 Collars A and B start from rest and move with the following accelerations: $a_A = 3t$ in./s^2 upward and $a_B = 9$ in./s^2 downward. Determine (a) the time at which the velocity of block C is again zero, (b) the distance through which block C will have moved at that time.

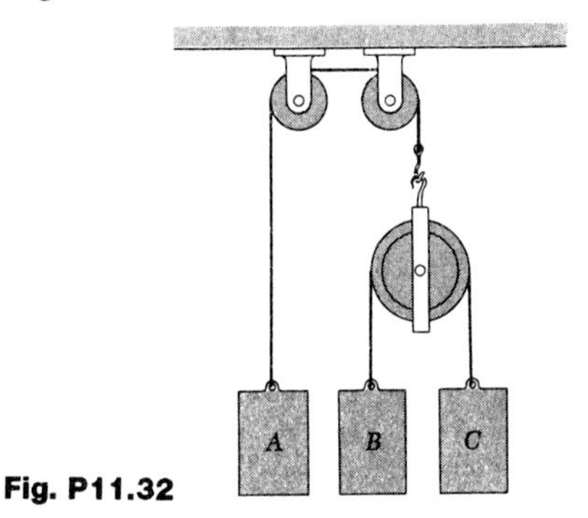

Fig. P11.32

11.32 (a) Choosing the positive sense downward for each block, express the velocity of A in terms of the velocities of B and C. (b) Knowing that both blocks A and C start from rest and move downward with the respective accelerations $a_A = 50$ mm/s^2 and $a_C = 110$ mm/s^2, determine the velocity and position of B after 3 s.

SECTIONS 11.7 to 11.8

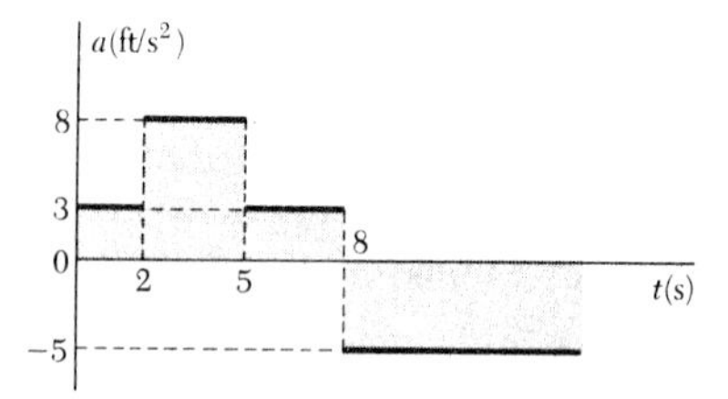

Fig. P11.33

11.33 A particle moves in a straight line with the acceleration shown in the figure. Knowing that it starts from the origin with $v_0 = -14$ ft/s, plot the v–t and x–t curves for $0 < t < 15$ s and determine (a) the maximum value of the velocity of the particle, (b) the maximum value of its position coordinate.

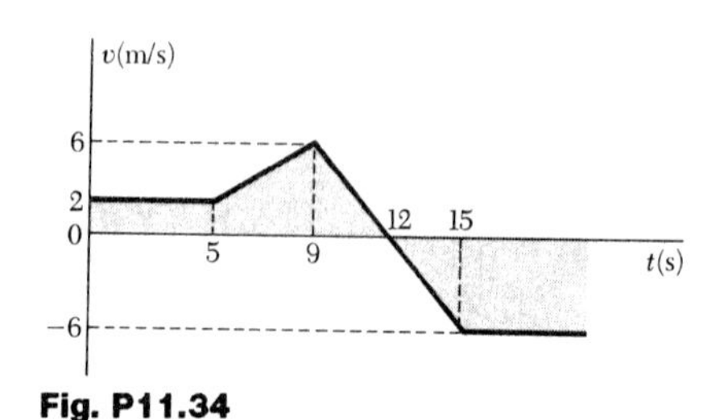

Fig. P11.34

11.34 A particle moves in a straight line with the velocity shown in the figure. Knowing that $x = -8$ m at $t = 0$, draw the a–t and x–t curves for $0 < t < 20$ s and determine (a) the maximum value of the position coordinate of the particle, (b) the values of t for which the particle is at a distance of 18 m from the origin.

11.35 A bus starts from rest at point A and accelerates at the rate of 0.75 m/s^2 until it reaches a speed of 9 m/s. It then proceeds at 9 m/s until the brakes are applied; it comes to rest at point B, 27 m beyond the point where the brakes were applied. Assuming uniform deceleration and knowing that the distance between A and B is 180 m, determine the time required for the bus to travel from A to B.

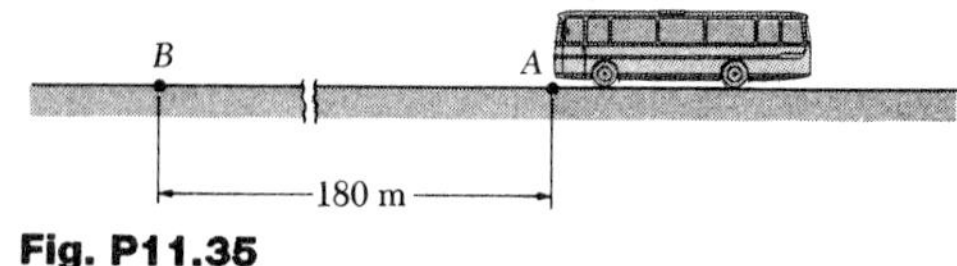

Fig. P11.35

11.36 Firing a howitzer causes the barrel to recoil 42 in. before a braking mechanism brings it to rest. From a high-speed photographic record, it is found that the maximum value of the recoil velocity is 250 in./s and that this is reached 0.02 s after firing. Assuming that the recoil period consists of two phases during which the acceleration has, respectively, a constant positive value a_1 and a constant negative value a_2, determine (*a*) the values of a_1 and a_2, (*b*) the position of the barrel 0.02 s after firing, (*c*) the time at which the velocity of the barrel is zero.

11.37 During a finishing operation the bed of an industrial planer moves alternately 36 in. to the right and 36 in. to the left. The velocity of the bed is limited to a maximum value of 6 in./s to the right and 9 in./s to the left; the acceleration is successively equal to 3 in./s^2 to the right, zero, 3 in./s^2 to the left, zero, etc. Determine the time required for the bed to complete a full cycle, and draw the v–t and x–t curves.

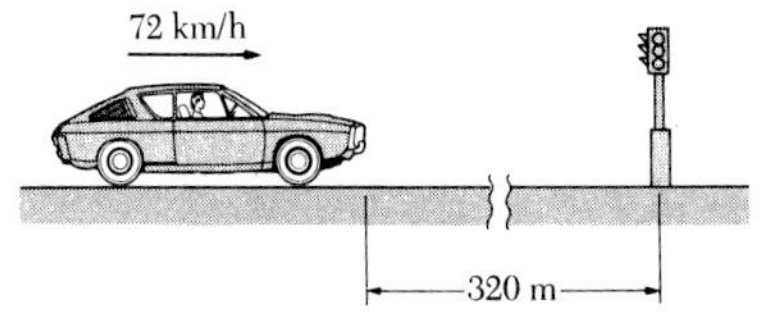

Fig. P11.38

11.38 A motorist is traveling at 72 km/h when she observes that a traffic signal 320 m ahead of her turns red. She knows that the signal is timed to stay red for 22 s. What should she do to pass the signal at 72 km/h just as it turns green again? Draw the v–t curve, selecting the solution which calls for the smallest possible deceleration and acceleration, and determine (*a*) the deceleration and acceleration in m/s^2, (*b*) the minimum speed reached in km/h.

11.39 An automobile at rest is passed by a truck traveling at a constant speed of 54 km/h. The automobile starts and accelerates for 10 s at a constant rate until it reaches a speed of 81 km/h. If the automobile then maintains a constant speed of 81 km/h, determine when and where it will overtake the truck, assuming that the automobile starts (*a*) just as the truck passes it, (*b*) 2 s after the truck has passed it.

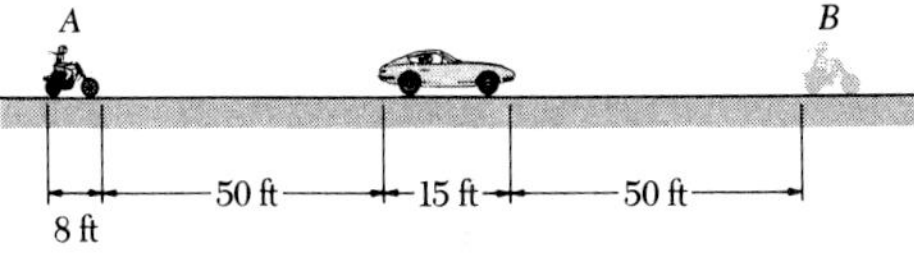

Fig. P11.40

11.40 A motorcycle and an automobile are both traveling at the constant speed of 40 mi/h; the motorcycle is 50 ft behind the automobile. The motorcyclist wants to pass the automobile, i.e., he wishes to place his motorcycle at *B*, 50 ft in front of the automobile, and then resume the speed of 40 mi/h. The maximum acceleration of the motorcycle is 6 ft/s^2 and the maximum deceleration obtained by applying the brakes is 18 ft/s^2. What is the shortest time in which the motorcyclist can complete the passing operation if he does not at any time exceed a speed of 55 mi/h? Draw the v–t curve.

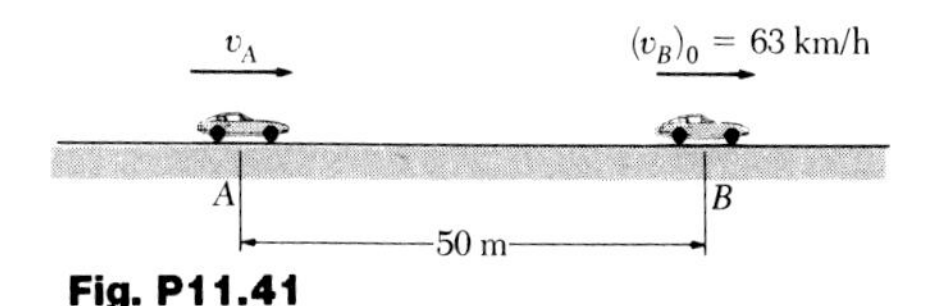

Fig. P11.41

11.41 Car *A* is traveling at the constant speed v_A. It approaches car *B*, which is traveling in the same direction at the constant speed of 63 km/h. The driver of car *B* notices car *A* when it is still 50 m behind him and then accelerates at the constant rate of 0.8 m/s^2 to avoid being passed or struck by car *A*. Knowing that the closest that *A* comes to *B* is 10 m, determine the speed v_A of car *A*.

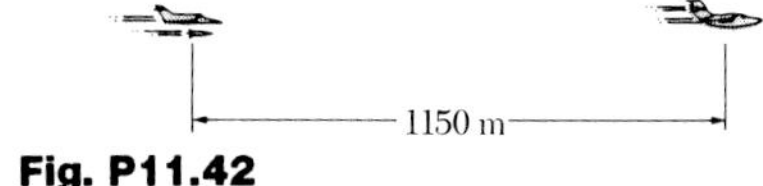

Fig. P11.42

11.42 A fighter plane flying horizontally in a straight line at 900 km/h is overtaking a bomber flying in the same straight line at 720 km/h. The pilot of the fighter plane fires an air-to-air missile at the bomber when his plane is 1150 m behind the bomber. The missile accelerates at a constant rate of 400 m/s^2 for 1 s and then travels at a constant speed. (*a*) How many seconds after firing will the missile reach the bomber? (*b*) If both planes continue at constant speeds, what will be the distance between the planes when the missile strikes the bomber?

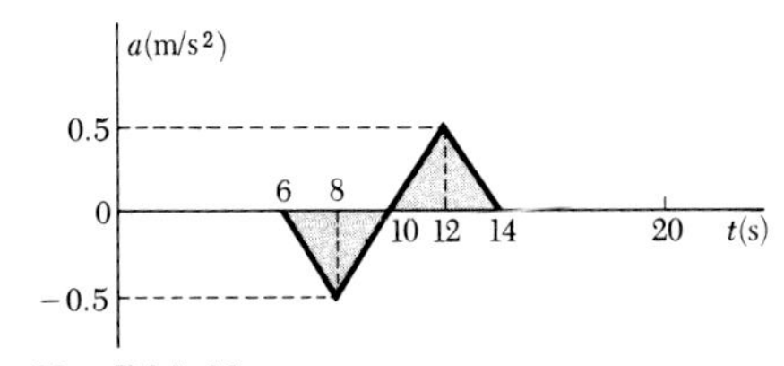

Fig. P11.43

11.43 The acceleration record shown was obtained for a small airplane traveling along a straight course. Knowing that $x = 0$ and $v = 50$ m/s when $t = 0$, determine (*a*) the velocity and position of the plane at $t = 20$ s, (*b*) its average velocity during the interval $6\text{ s} < t < 14\text{ s}$.

11.44 A train starts at a station and accelerates uniformly at a rate of 1 ft/s^2 until it reaches a speed of 20 ft/s; it then proceeds at the constant speed of 20 ft/s. Determine the time and the distance traveled if its *average* velocity is (*a*) 12 ft/s, (*b*) 18 ft/s.

11.45 The rate of change of acceleration is known as the *jerk;* large or abrupt rates of change of acceleration cause discomfort to elevator passengers. If the jerk, or rate of change of the acceleration, of an elevator is limited to ± 1.5 ft/s^2 per second, determine (*a*) the shortest time required for an elevator, starting from rest, to rise 24 ft and stop, (*b*) the corresponding average velocity of the elevator.

11.46 In order to maintain passenger comfort, the acceleration of an elevator is limited to ± 1.2 m/s^2 and the jerk, or rate of change of acceleration, is limited to ± 0.4 m/s^2 per second. If the elevator starts from rest, determine (*a*) the shortest time required for it to attain a constant velocity of 6 m/s, (*b*) the distance traveled in that time, (*c*) the corresponding average velocity of the elevator.

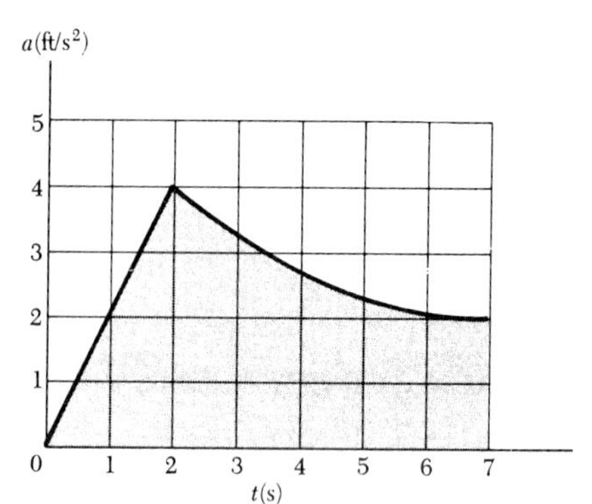

Fig. P11.47

11.47 The acceleration record shown was obtained for an automobile traveling on a straight highway. Knowing that the initial velocity of the automobile was 15 mi/h, determine the velocity and distance traveled when (*a*) $t = 3$ s, (*b*) $t = 6$ s.

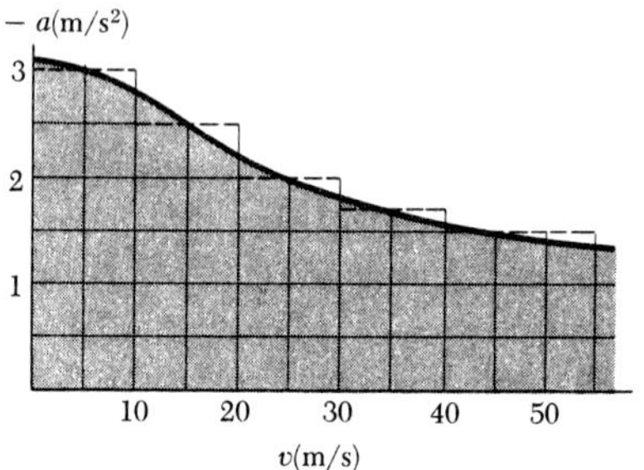

Fig. P11.48

11.48 The maximum possible deceleration of a passenger train under emergency conditions was determined experimentally; the results are shown (solid curve) in the figure. If the brakes are applied when the train is traveling at 108 km/h, determine by approximate means (*a*) the time required for the train to come to rest, (*b*) the distance traveled in that time.

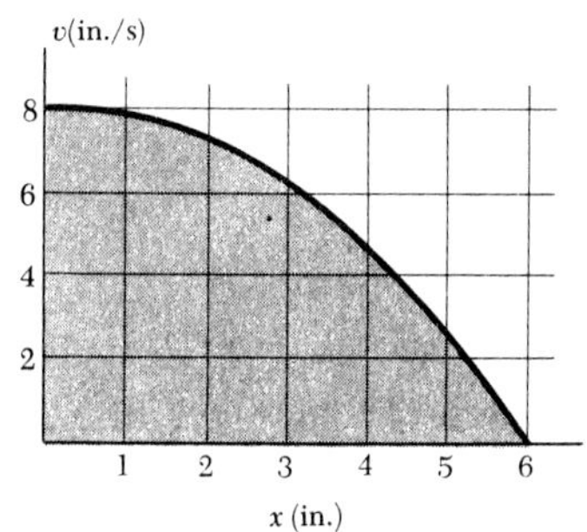

Fig. P11.49

11.49 The v–x curve shown was obtained experimentally during the motion of the bed of an industrial planer. Determine by approximate means the acceleration (*a*) when $x = 3$ in., (*b*) when $v = 4$ in./s.

SECTION 11.9 to 11.12

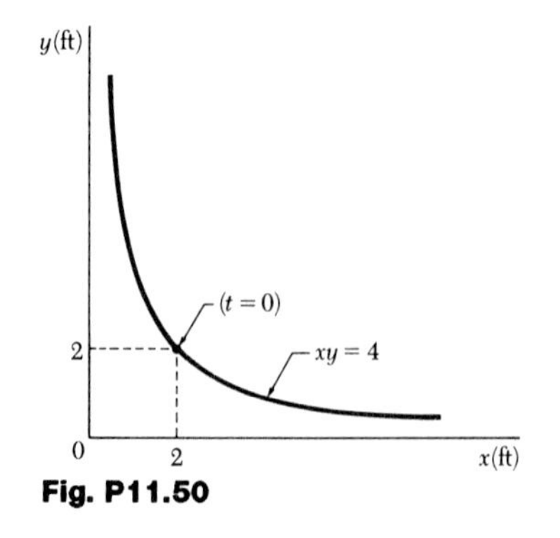

Fig. P11.50

11.50 The motion of a particle is defined by the equations $x = 2(t + 1)^2$ and $y = 2(t + 1)^{-2}$, where x and y are expressed in feet and t in seconds. Show that the path of the particle is part of the rectangular hyperbola shown and determine the velocity and acceleration when (*a*) $t = 0$, (*b*) $t = \frac{1}{2}$ s.

11.51 The motion of a particle is defined by the equations $x = 2t^2 - 4t$ and $y = 2(t - 1)^2 - 4(t - 1)$, where x and y are expressed in meters and t in seconds. Determine (*a*) the magnitude of the smallest velocity reached by the particle, (*b*) the corresponding time, position, and direction of the velocity.

11.52 A particle moves in an elliptic path defined by the position vector $\mathbf{r} = (A \cos pt)\mathbf{i} + (B \sin pt)\mathbf{j}$. Show that the acceleration (*a*) is directed toward the origin, (*b*) is proportional to the distance from the origin to the particle.

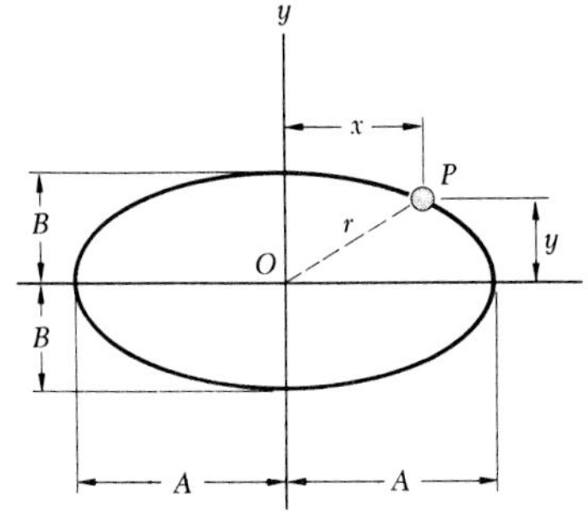

Fig. P11.52

11.53 A man standing at the 18-m level of a tower throws a stone in a horizontal direction. Knowing that the stone hits the ground 25 m from the bottom of the tower, determine (*a*) the initial velocity of the stone, (*b*) the distance at which a stone would hit the ground if it were thrown horizontally with the same velocity from the 22-m level of the tower.

11.54 A handball player throws a ball from *A* with a horizontal velocity $\mathbf{v}_0$. Knowing that $d = 20$ ft, determine (*a*) the value of v_0 for which the ball will strike the corner *C*, (*b*) the range of values of v_0 for which the ball will strike the corner region *BCD*.

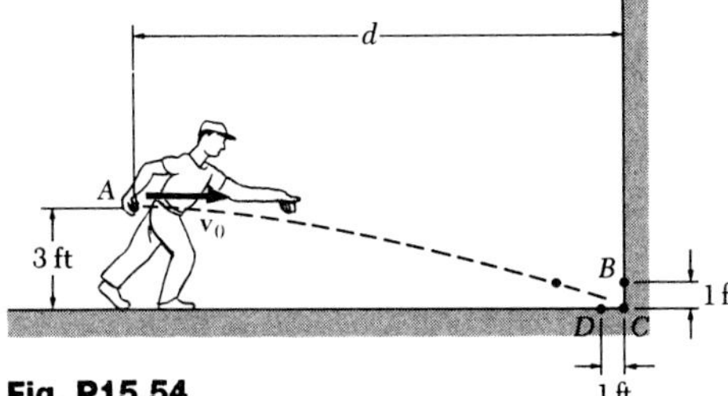

Fig. P15.54

11.55 Sand is discharged at *A* from a horizontal conveyor belt with an initial velocity $\mathbf{v}_0$. Determine the range of values of v_0 for which the sand will enter the vertical chute shown.

11.56 A pump is located near the edge of the horizontal platform shown. The nozzle at *A* discharges water with an initial velocity of 25 ft/s at an angle of 50° with the vertical. Determine the range of values of the height *h* for which the water enters the opening *BC*.

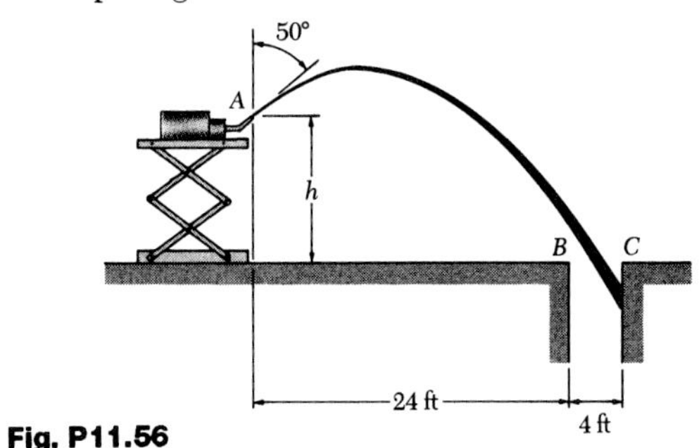

Fig. P11.56

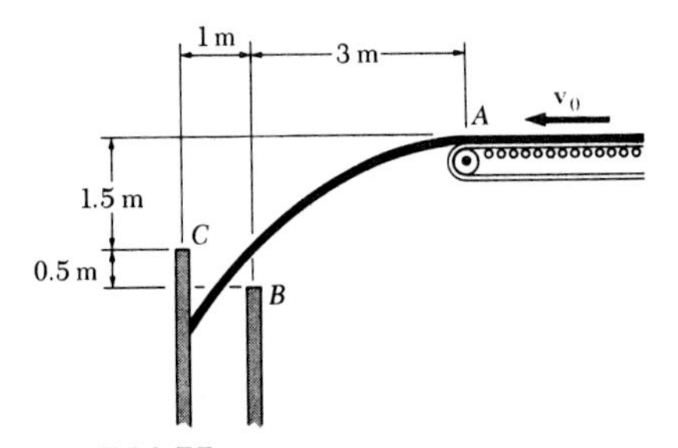

Fig. P11.55

11.57 An oscillating water sprinkler is operated at point *A* on an incline which forms an angle α with the horizontal. The sprinkler discharges water with an initial velocity $\mathbf{v}_0$ at an angle ϕ with the vertical which varies from $-\phi_0$ to $+\phi_0$. Knowing that $v_0 = 10$ m/s, $\phi_0 = 40°$, and $\alpha = 10°$, determine the horizontal distance between the sprinkler and points *B* and *C* which define the watered area.

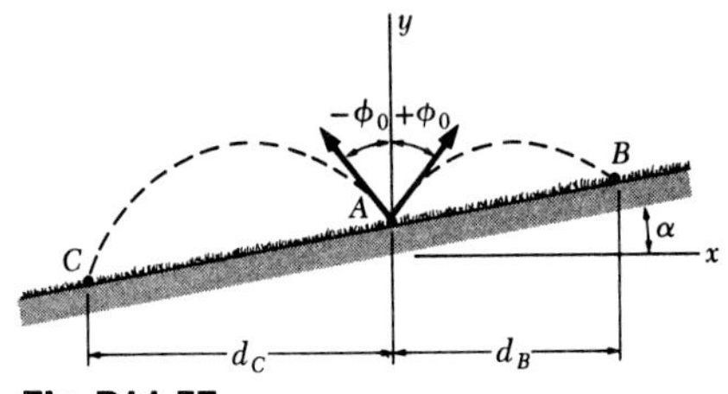

Fig. P11.57

11.58 A nozzle at *A* discharges water with an initial velocity of 12 m/s at an angle of 60° with the horizontal. Determine where the stream of water strikes the roof. Check that the stream will clear the edge of the roof.

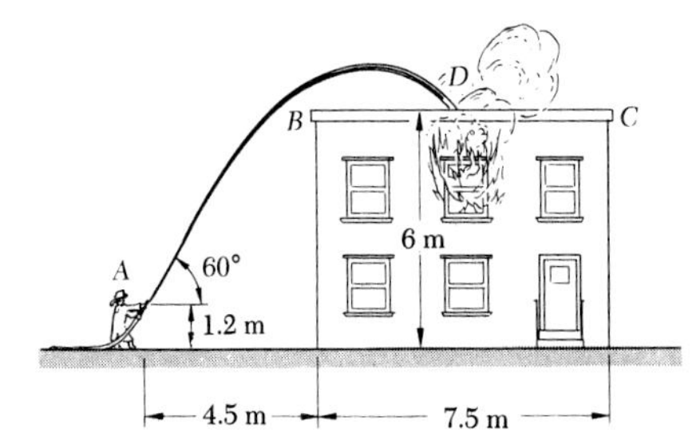

Fig. P11.58

11.59 A player throws a ball with an initial velocity $\mathbf{v}_0$ of 15 m/s from a point *A* located 1.5 m above the floor. Knowing that $h = 3$ m, determine the angle α for which the ball will strike the wall at point *B*.

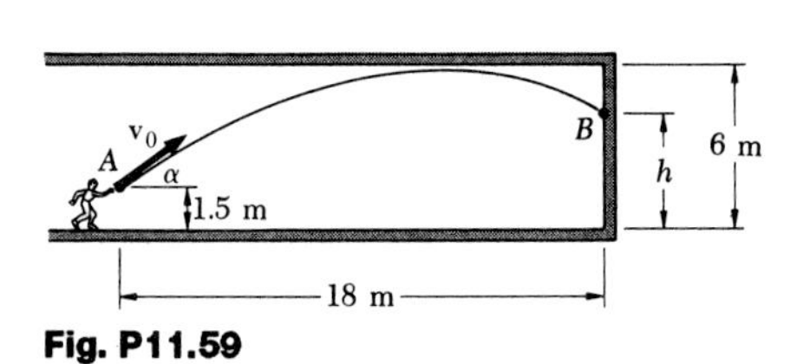

Fig. P11.59

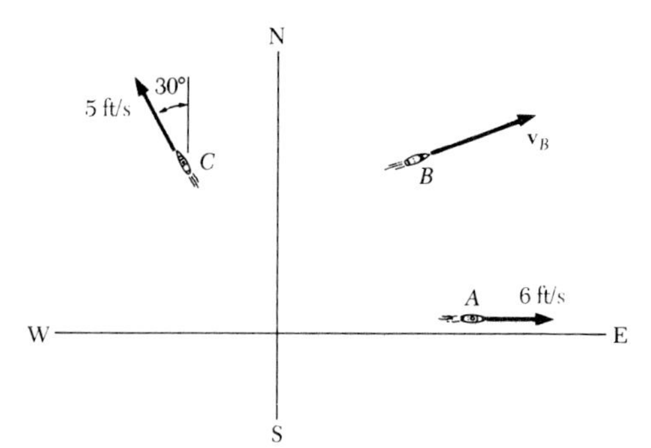

Fig. P11.60

11.60 The velocities of boats A and C are as shown and the relative velocity of boat B with respect to A is $\mathbf{v}_{B/A} = 4$ m/s $\measuredangle$ 50°. Determine (*a*) $v_{A/C}$, (*b*) $v_{C/B}$, (*c*) the change in position of B with respect to C during a 10-s interval. Also show that for any motion, $\mathbf{v}_{B/A} + \mathbf{v}_{C/B} + \mathbf{v}_{A/C} = 0$.

11.61 Instruments in an airplane indicate that with respect to the air, the plane is moving north at a speed of 500 km/h. At the same time ground-based radar indicates that the plane is moving at a speed of 530 km/h in a direction 5° east of north. Determine the magnitude and direction of the velocity of the air.

11.62 Two airplanes A and B are flying at the same altitude; plane A is flying due east at a constant speed of 900 km/h, while plane B is flying southwest at a constant speed of 600 km/h. Determine the change in position of plane B relative to plane A, which takes place during a 2-min interval.

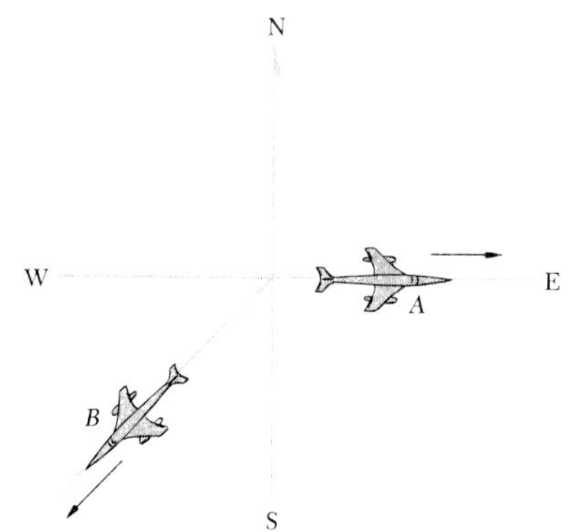

Fig. P11.62

11.63 Four seconds after automobile B passes through the intersection shown, automobile A passes through the same intersection. Knowing that the speed of each automobile is constant, determine (*a*) the relative velocity of B with respect to A, (*b*) the change in position of B with respect to A during a 3-s interval, (*c*) the distance between the two automobiles 2 s after A has passed through the intersection.

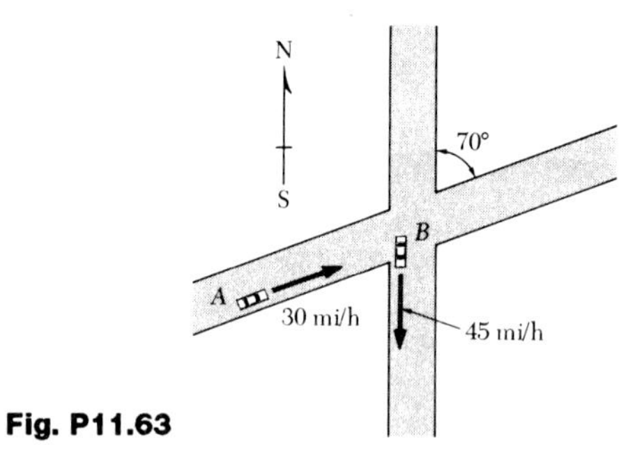

Fig. P11.63

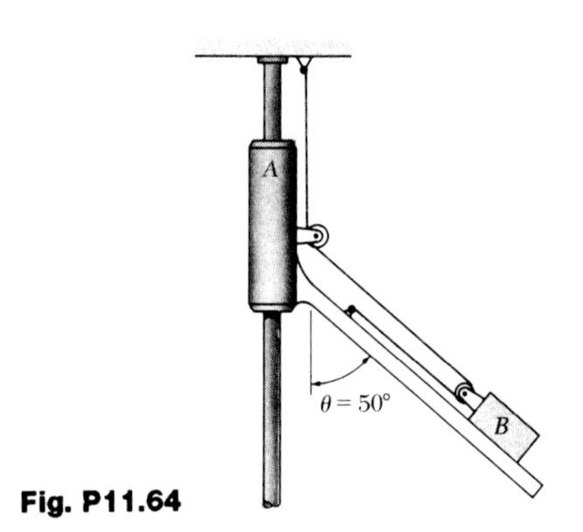

Fig. P11.64

11.64 Knowing that at the instant shown assembly A has a velocity of 16 in./s and an acceleration of 24 in./s^2 both directed downward, determine (*a*) the velocity of block B, (*b*) the acceleration of block B.

11.65 An antiaircraft gun fires a shell as a plane passes directly over the position of the gun, at an altitude of 6000 ft. The muzzle velocity of the shell is 1500 ft/s. Knowing that the plane is flying horizontally at 450 mi/h, determine (*a*) the required firing angle if the shell is to hit the plane, (*b*) the velocity and acceleration of the shell relative to the plane at the time of impact.

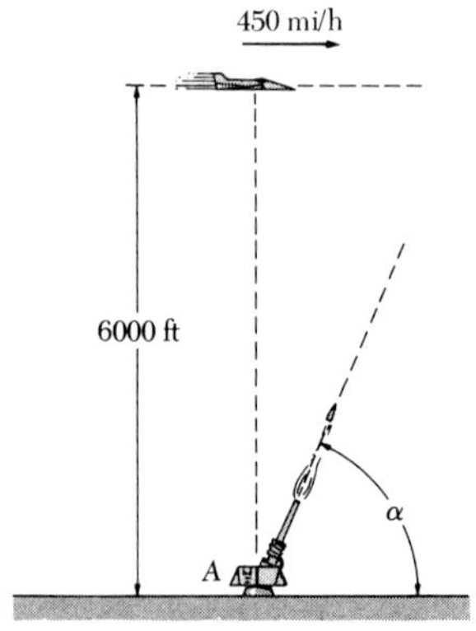

Fig. P11.65

11.66 What is the smallest radius which should be used for a highway curve if the normal component of the acceleration of a car traveling at 60 mi/h is not to exceed 2.5 ft/s^2?

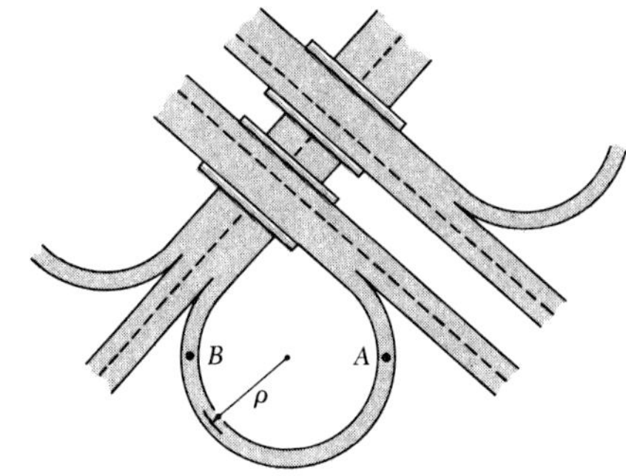

Fig. P11.67

11.67 A motorist drives along the circular exit ramp of a turnpike at the constant speed v_0. Knowing that the odometer indicates a distance of 0.6 km between point A where the automobile is going due south and B where it is going due north, determine the speed v_0 for which the normal component of the acceleration is $0.08g$.

11.68 A computer tape moves over two drums at a constant speed v_0. Knowing that the normal component of the acceleration of the portion of tape in contact with drum B is 400 ft/s^2, determine (a) the speed v_0, (b) the normal component of the acceleration of the portion of tape in contact with drum A.

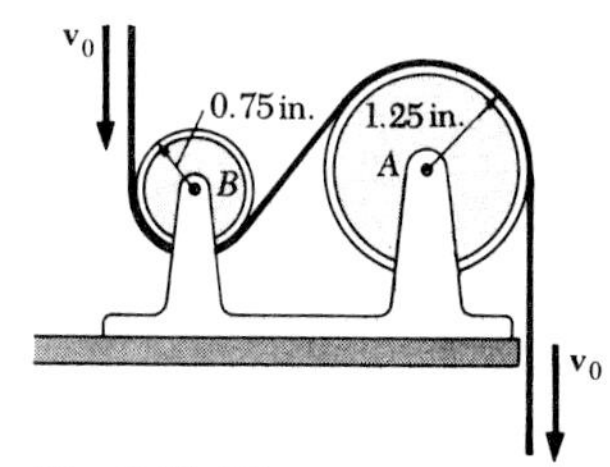

Fig. P11.68

11.69 A motorist is traveling on a curved portion of highway of radius 400 m at a speed of 90 km/h. The brakes are suddenly applied, causing the speed to decrease at a constant rate of 1.2 m/s^2. Determine the magnitude of the total acceleration of the automobile (a) immediately after the brakes have been applied, (b) 5 s later.

11.70 A bus starts from rest on a curve of 250-m radius and accelerates at the constant rate $a_t = 0.6$ m/s^2. Determine the distance and time that the bus will travel before the magnitude of its total acceleration is 0.75 m/s^2.

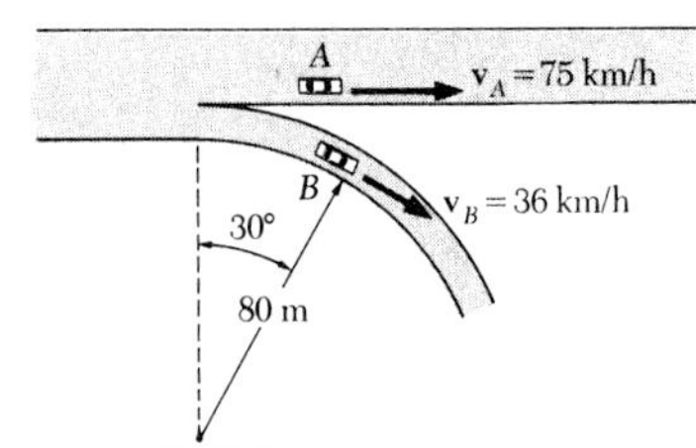

Fig. P11.72

11.71 A motorist decreases the speed of an automobile at a constant rate from 45 to 30 mi/h over a distance of 750 ft along a curve of 1500-ft radius. Determine the magnitude of the total acceleration of the automobile after it has traveled 500 ft along the curve.

11.72 Automobile A is traveling along a straight highway, while B is moving along a circular exit ramp of 80-m radius. The speed of A is being increased at the rate of 2 m/s^2 and the speed of B is being decreased at the rate of 1.2 m/s^2. For the position shown, determine (a) the velocity of A relative to B, (b) the acceleration of A relative to B.

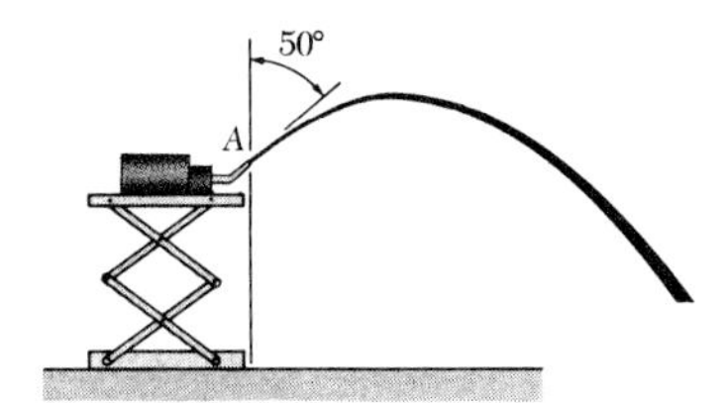

Fig. P11.73

11.73 A nozzle discharges a stream of water in the direction shown with an initial velocity of 25 ft/s. Determine the radius of curvature of the stream (a) as it leaves the nozzle, (b) at the maximum height of the stream.

11.74 A satellite will travel indefinitely in a circular orbit around the earth if the normal component of its acceleration is equal to $g(R/r)^2$, where $g = 32.2$ ft/s^2, R = radius of the earth = 3960 mi, and r = distance from the center of the earth to the satellite. Determine the height above the surface of the earth at which a satellite will travel indefinitely around the earth at a speed of 16,500 mi/h.

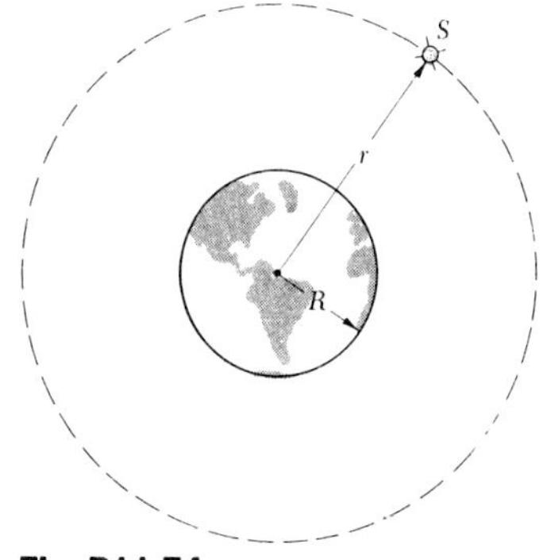

Fig. P11.74

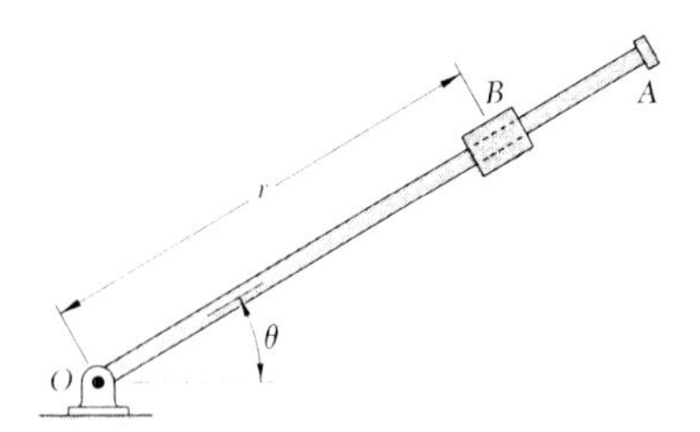

Fig. P11.75

11.75 The rotation of rod OA about O is defined by the relation $\theta = t^3 - 4t$, where θ is expressed in radians and t in seconds. Collar B slides along the rod in such a way that its distance from O is $r = 25t^3 - 50t^2$, where r is expressed in millimeters and t in seconds. When $t = 1$ s, determine (*a*) the velocity of the collar, (*b*) the total acceleration of the collar, (*c*) the acceleration of the collar relative to the rod.

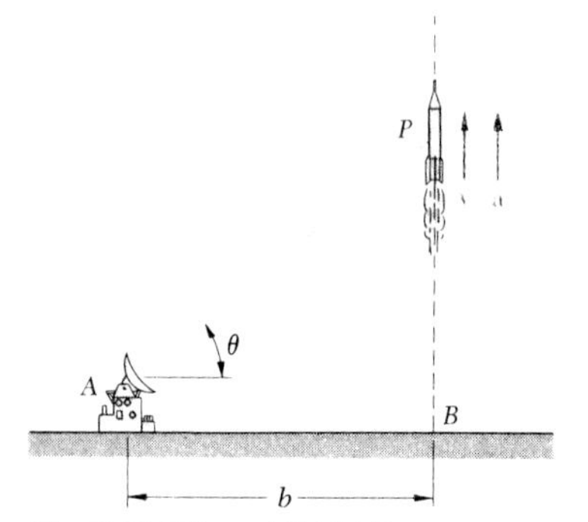

Fig. P11.76 and P11.78

11.76 A rocket is fired vertically from a launching pad at B. Its flight is tracked by radar from point A. Determine the velocity of the rocket in terms of b, θ, and $\dot{\theta}$.

11.77 The flight path of airplane B is a horizontal straight line and passes directly over a radar tracking station at A. Knowing that the airplane moves to the left with the constant velocity $\mathbf{v}_0$, determine $d\theta/dt$ in terms of v_0, h, and θ.

11.78 A test rocket is fired vertically from a launching pad at B. When the rocket is at P the angle of elevation is $\theta = 42.0°$, and 0.5 s later it is $\theta = 43.2°$. Knowing that $b = 3$ km, determine approximately the speed of the rocket during the 0.5-s interval.

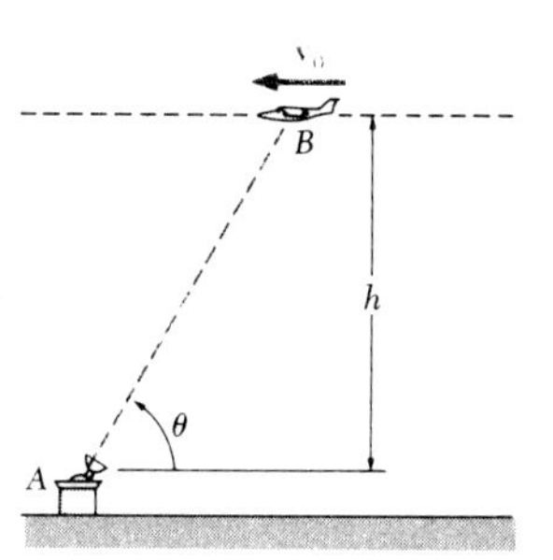

Fig. P11.77

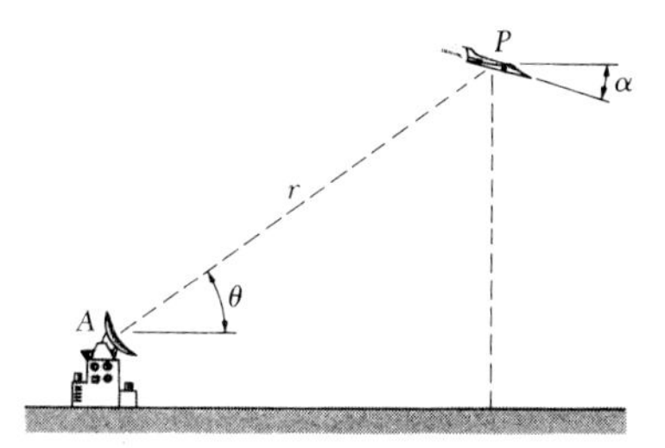

Fig. P11.79

11.79 An airplane passes over a radar tracking station at A and continues to fly due east. When the airplane is at P, the distance and angle of elevation of the plane are, respectively, $r = 11{,}200$ ft and $\theta = 26.5°$. Two seconds later the radar station sights the plane at $r = 12{,}300$ ft and $\theta = 23.3°$. Determine approximately the speed and the angle of dive α of the plane during the 2-s interval.

11.80 As rod OA rotates, pin P moves along the parabola BCD. Knowing that the equation of the parabola is $r = 2b/(1 + \cos\theta)$ and that $\theta = kt$, determine the velocity and acceleration of P when (*a*) $\theta = 0$, (*b*) $\theta = 90°$.

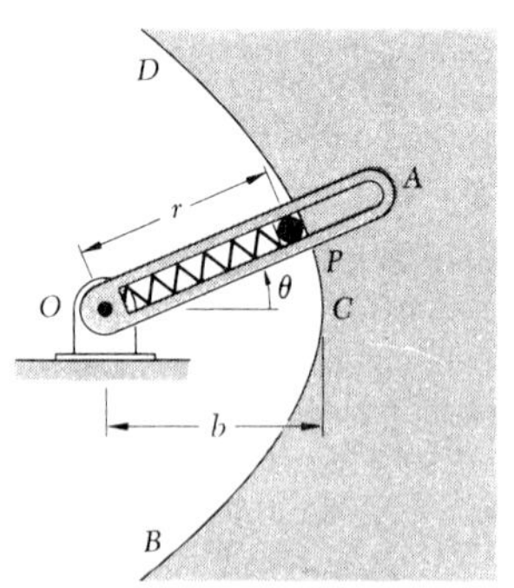

Fig. P11.80

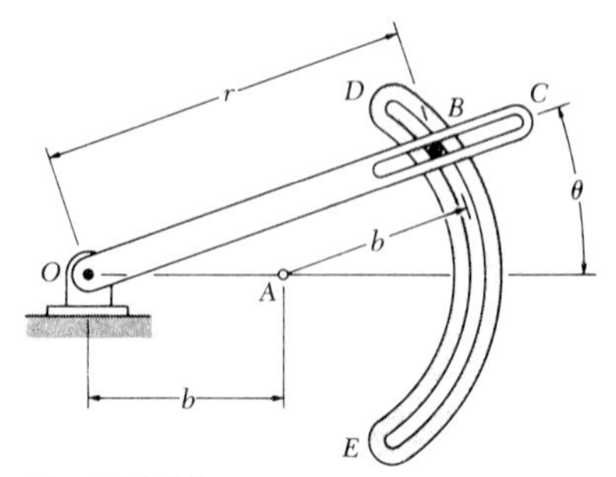

Fig. P11.81

11.81 The pin at B is free to slide along the circular slot DE and along the rotating rod OC. Assuming that the rod OC rotates at a constant rate $\dot{\theta}$, (*a*) show that the acceleration of pin B is of constant magnitude, (*b*) determine the direction of the acceleration of pin B.

CHAPTER 11 COMPUTER PROBLEMS

11.C1 The magnitude in m/s^2 of the deceleration due to air resistance of the nose cone of a small experimental rocket is known to be $0.0005v^2$, where v is expressed in m/s. The nose cone is projected vertically from the ground with an initial velocity of 100 m/s. Derive expressions for the velocity of the cone as a function of height as the cone moves up to its maximum height and then back down to the ground. Plot the velocity as a function of the height. [Hint: The total acceleration of the cone as it moves up is $-(g + 0.0005v^2)$, and as it moves down is $-(g - 0.0005v^2)$, where g = 9.81 m/s^2.]

11.C2 A projectile enters a resisting medium at $x = 0$ with an initial velocity $v_0 = 900$ ft/s. The velocity of the projectile is defined by the relation $v = v_0 - kx$, where v is expressed in ft/s, x in feet, and $k = 2700$ s^{-1}. Derive expressions for the position, velocity, and acceleration of the projectile as functions of time. Plot the position, velocity, and acceleration of the projectile as functions of time from the time the projectile enters the medium.

11.C3 The motion of a particle is defined by the equations $x = 1.5t^2 - 6t$ and $y = 6t^2 - 2t^3$, where x and y are expressed in inches and t in seconds. Derive expressions for the magnitudes of the velocity and acceleration of the particle as a function of time.

(d) Plot the trajectory of the particle.
(e) Plot the rectangular components of the velocity v_x and v_y and the total velocity v as a function of time for t =0 until t = 20 s.
(f) Plot the rectangular components of the acceleration a_x and a_y and the total acceleration a as a function of time for t =0 until t = 20 s.

11.C4 The rotation of rod OA about O is defined by the relation $\theta = t^3 + 4t$, where θ is expressed in radians and t in seconds. Collar B slides along the rod in such a way that its distance from O is $r = 25t^3 + 50t^2$, where r is expressed in millimeters and t in seconds. Plot the position of the particle for one revolution of the rod starting from $t = 0$. Derive expressions for the radial and transverse components of the velocity and acceleration as a function of time. Plot the radial and transverse components of the velocity and the magnitude of the velocity as functions of time for one revolution of the rod. Plot the radial and transverse components of the acceleration and the magnitude of the acceleration as functions of time for one revolution of the rod.

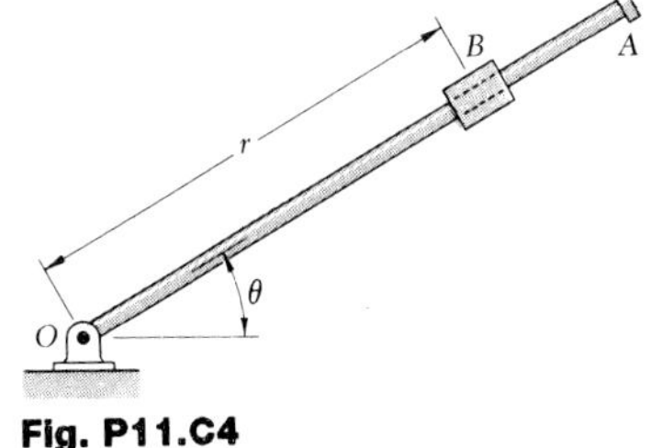

Fig. P11.C4

11.C5 The motion of a particle along an elliptical path is defined by the relations $r = 35/(1 - 0.75\cos \pi t)$ and $\theta = \pi t$, where r is expressed in millimeters and θ in radians. Plot the position of the particle for $\theta = 0$ to $\theta = 2\pi$. Derive expressions for the radial and transverse components of the velocity and acceleration as functions of time. Plot the radial and transverse components of the velocity and the magnitude of the velocity as functions of time for t = 0 to t = 2 s. Plot the radial and transverse components of the acceleration and the magnitude of the acceleration as functions of time for t = 0 to t = 2 s.

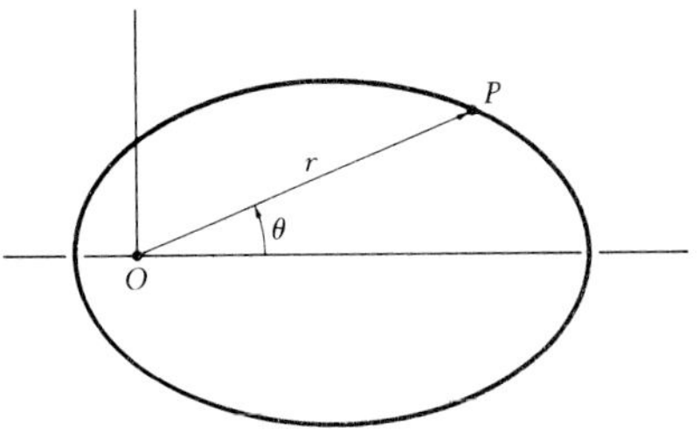

Fig. P11.C5

CHAPTER 12
KINEMATICS OF PARTICLES: NEWTON'S SECOND LAW

SECTIONS 12.1 to 12.6

12.1 A motorist traveling at a speed of 90 km/h suddenly applies the brakes and comes to a stop after skidding 50 m. Determine (*a*) the time required for the car to stop, (*b*) the coefficient of friction between the tires and the pavement.

12.2 A car has been traveling up a long 2 percent grade at a constant speed of 55 mi/h. If the driver does not change the setting of the throttle or shift gears as the car reaches the top of the hill, what will be the acceleration of the car as it starts moving down the 3 percent grade?

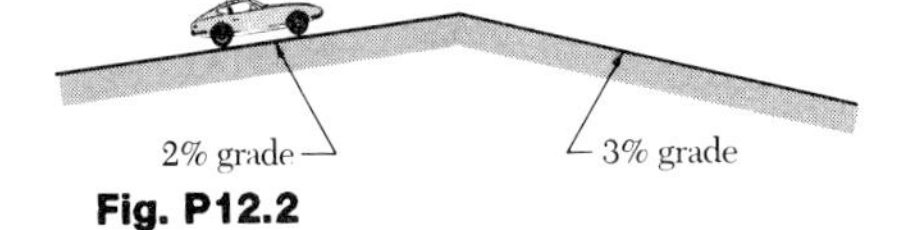

Fig. P12.2

12.3 The two blocks shown are originally at rest. Neglecting the masses of the pulleys and the effect of friction in the pulleys and between block A and the incline, determine (*a*) the acceleration of each block, (*b*) the tension in the cable.

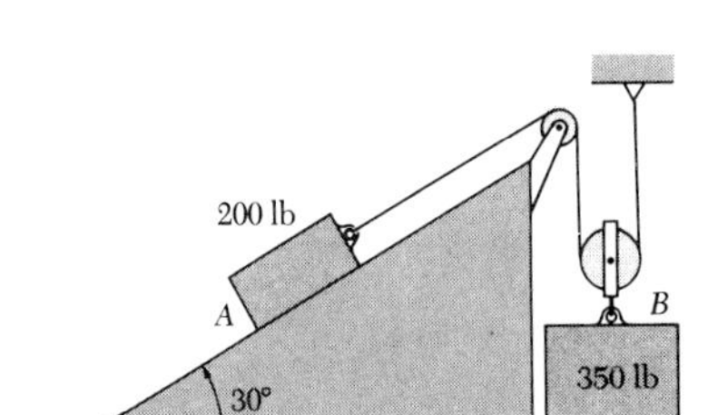

Fig. P12.3

12.4 A 1050-kg trailer is hitched to a 1200-kg car. The car and trailer are traveling at 90 km/h when the driver applies the brakes on both the car and the trailer. Knowing that the braking forces exerted on the car and the trailer are 4500 N and 3600 N, respectively, determine (*a*) the deceleration of the car and trailer, (*b*) the horizontal component of the force exerted by the trailer hitch on the car.

Fig. P12.4

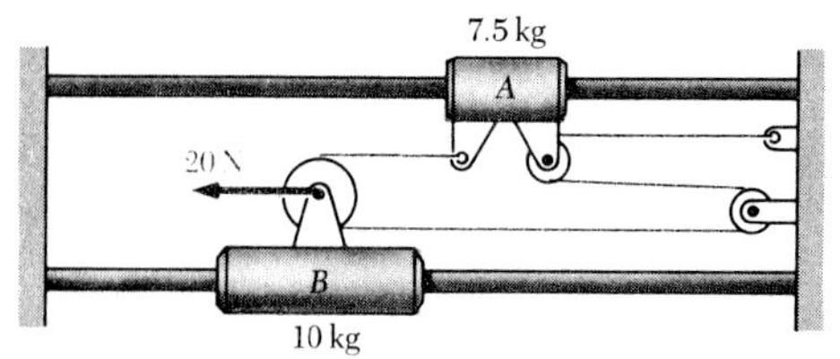

Fig. P12.5

12.5 Knowing that the system shown starts from rest, find the velocity at $t = 1.2$ s of (a) collar A, (b) collar B. Neglect the masses of the pulleys and the effect of friction.

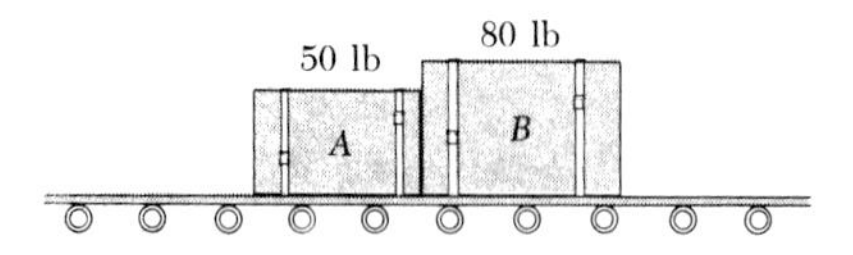

Fig. P12.6

12.6 Two packages are placed on a conveyor belt which is at rest. The coefficient of kinetic friction is 0.20 between the belt and package A, and 0.10 between the belt and package B. If the belt is suddenly started to the right and slipping occurs between the belt and the packages, determine (a) the acceleration of the packages, (b) the force exerted by package A on package B.

12.7 The coefficients of friction between the load and the flat-bed trailer shown are $\mu_s = 0.50$ and $\mu_k = 0.40$. Knowing that the speed of the rig is 45 mi/h, determine the shortest distance in which the rig can be brought to a stop if the load is not to shift.

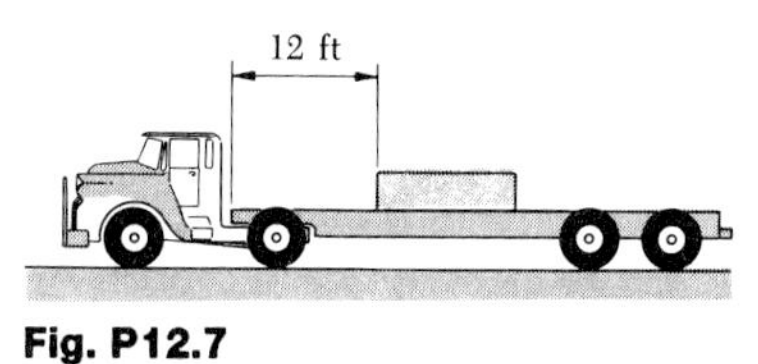

Fig. P12.7

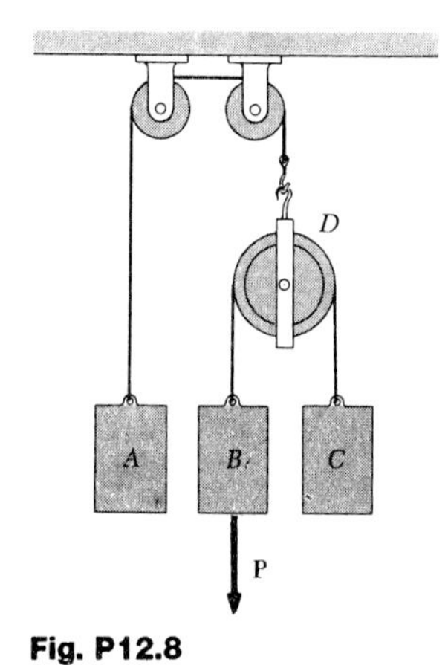

Fig. P12.8

12.8 Block A weighs 20 lb, and blocks B and C weigh 10 lb each. Knowing that $P = 2.5$ lb and that the blocks are initially at rest, determine at $t = 3$ s the velocity (a) of B relative to A, (b) of C relative to A. Neglect the weights of the pulleys and the effect of friction.

12.9 An 8-kg block B rests as shown on a 12-kg bracket A. The coefficients of friction are $\mu_s = 0.40$ and $\mu_k = 0.30$ between block B and bracket A, and there is no friction in the pulley or between the bracket and the horizontal surface. If $P = 30$ N, determine the velocity of B relative to A after B has moved 400 mm with respect to A.

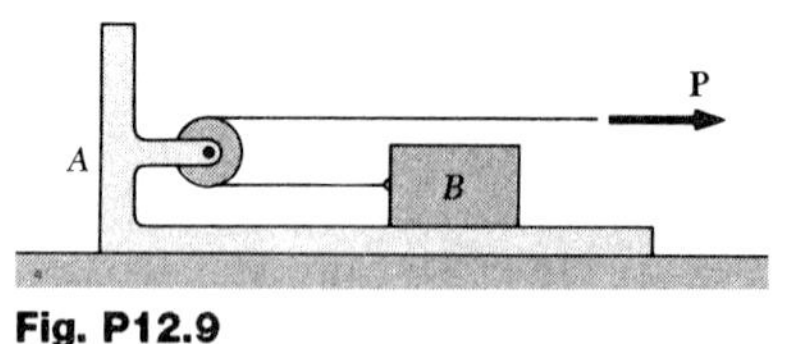

Fig. P12.9

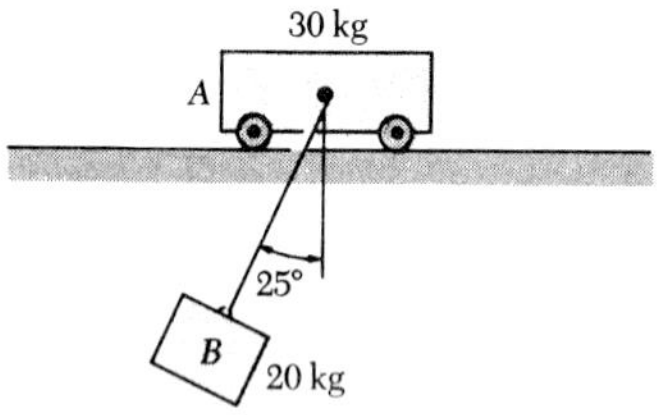

Fig. P12.10

12.10 A 20-kg block B is suspended from a 2-m cord attached to a 30-kg cart A. Neglecting friction, determine (a) the acceleration of the cart, (b) the tension in the cord, immediately after the system is released from rest in the position shown.

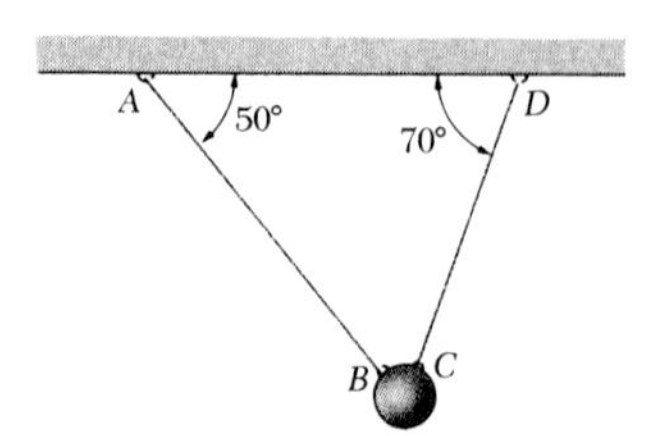

Fig. P12.11

12.11 A small sphere of weight W is held as shown by two wires AB and CD. If wire AB is cut, determine the tension in the other wire (a) before AB is cut, (b) immediately after AB has been cut.

12.12 A stunt driver proposes to drive a small automobile at the speed of 40 mi/h on the vertical wall of a circular pit of radius 50 ft. Knowing that the mass center of the automobile and driver is 2 ft from the wall, determine the minimum required value of the coefficient of static friction between the tires and the wall.

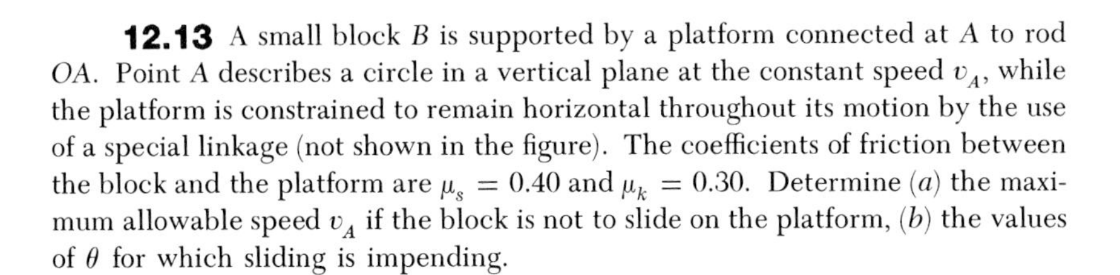

12.13 A small block B is supported by a platform connected at A to rod OA. Point A describes a circle in a vertical plane at the constant speed v_A, while the platform is constrained to remain horizontal throughout its motion by the use of a special linkage (not shown in the figure). The coefficients of friction between the block and the platform are $\mu_s = 0.40$ and $\mu_k = 0.30$. Determine (a) the maximum allowable speed v_A if the block is not to slide on the platform, (b) the values of θ for which sliding is impending.

Fig. P12.13

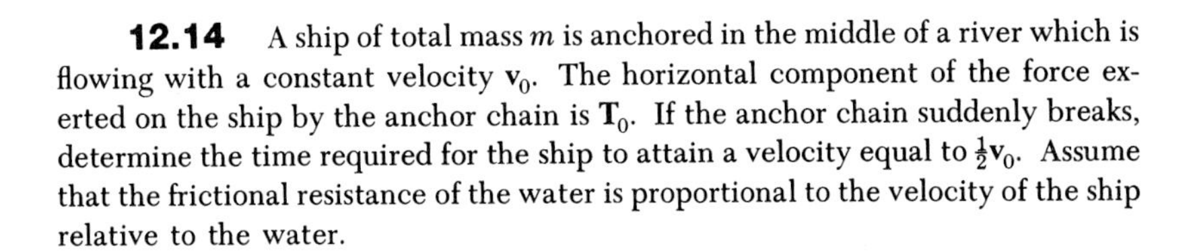

12.14 A ship of total mass m is anchored in the middle of a river which is flowing with a constant velocity $\mathbf{v}_0$. The horizontal component of the force exerted on the ship by the anchor chain is $\mathbf{T}_0$. If the anchor chain suddenly breaks, determine the time required for the ship to attain a velocity equal to $\frac{1}{2}\mathbf{v}_0$. Assume that the frictional resistance of the water is proportional to the velocity of the ship relative to the water.

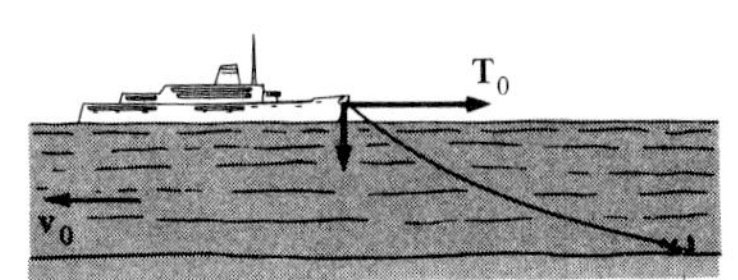

Fig. P12.14

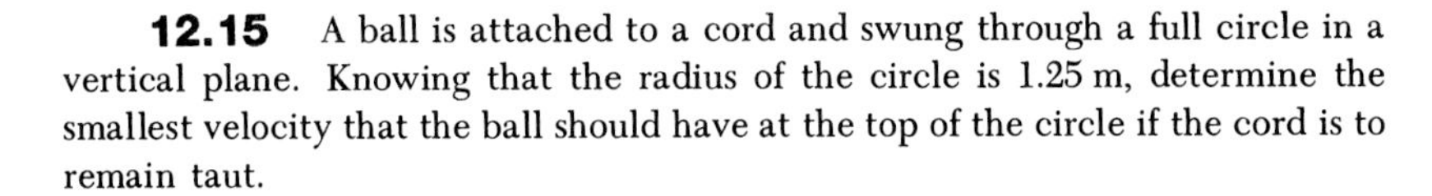

12.15 A ball is attached to a cord and swung through a full circle in a vertical plane. Knowing that the radius of the circle is 1.25 m, determine the smallest velocity that the ball should have at the top of the circle if the cord is to remain taut.

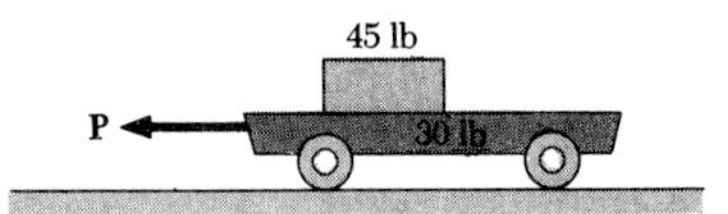

Fig. P12.16

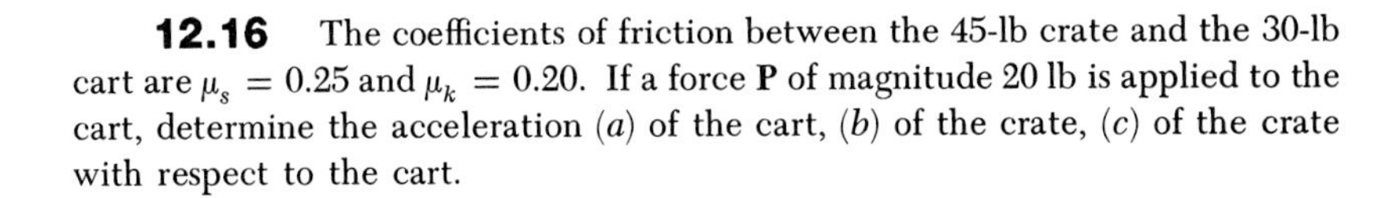

12.16 The coefficients of friction between the 45-lb crate and the 30-lb cart are $\mu_s = 0.25$ and $\mu_k = 0.20$. If a force $\mathbf{P}$ of magnitude 20 lb is applied to the cart, determine the acceleration (a) of the cart, (b) of the crate, (c) of the crate with respect to the cart.

12.17 The assembly shown rotates about a vertical axis at a constant rate. Knowing that the coefficient of static friction between the small block A and the cylindrical wall is 0.30, determine the lowest speed v for which the block will remain in contact with the wall.

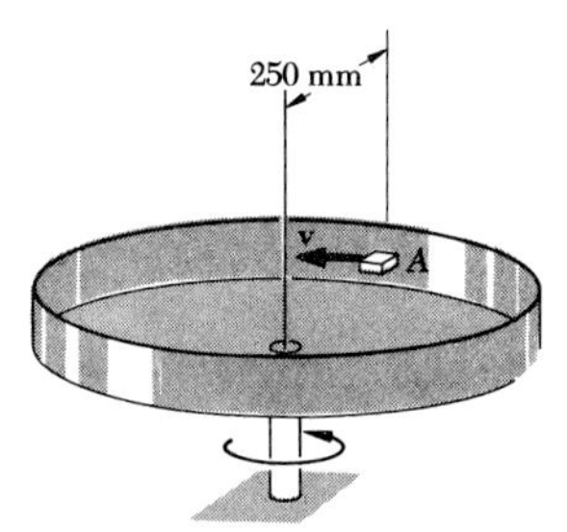

Fig. P12.17

12.18 Knowing that blocks B and C strike the ground simultaneously and exactly 1 s after the system is released from rest, determine m_B and m_C in terms of m_A.

12.19 Determine the acceleration of each block when $m_A = 5$ kg, $m_B = 15$ kg, and $m_C = 10$ kg. Which block strikes the ground first?

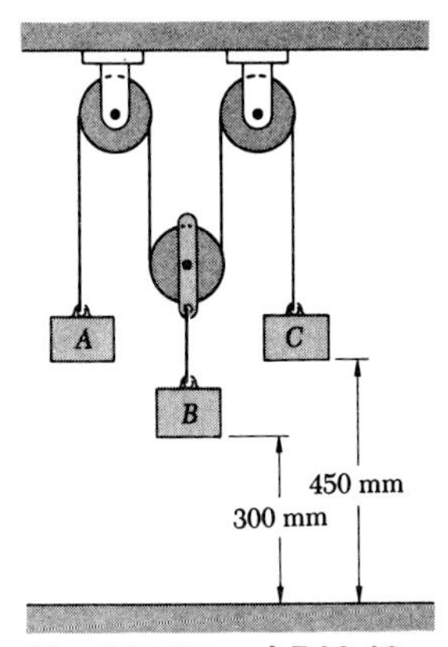

Fig. P12.18 and P12.19

12.20 Neglecting the effect of friction, determine (a) the acceleration of each block, (b) the tension in the cable.

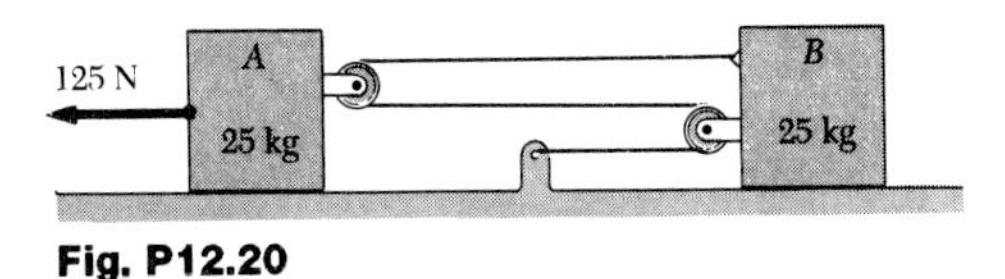

Fig. P12.20

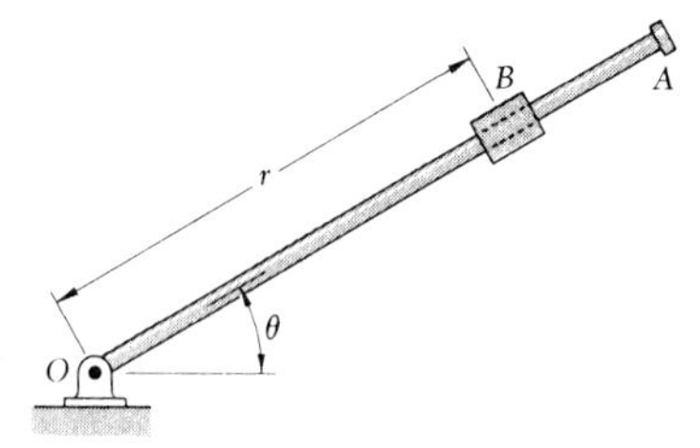

Fig. P12.21

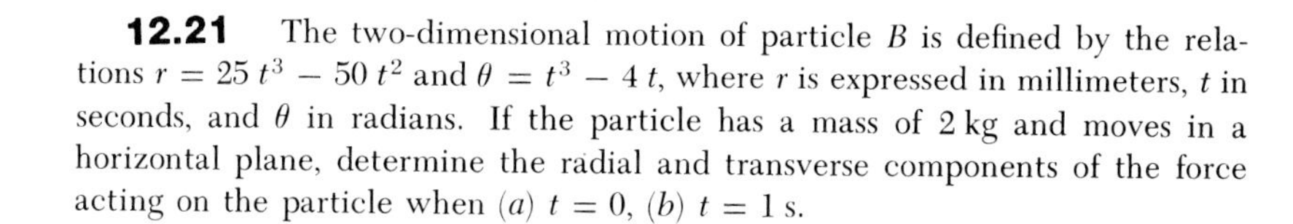
12.21 The two-dimensional motion of particle B is defined by the relations $r = 25\,t^3 - 50\,t^2$ and $\theta = t^3 - 4\,t$, where r is expressed in millimeters, t in seconds, and θ in radians. If the particle has a mass of 2 kg and moves in a horizontal plane, determine the radial and transverse components of the force acting on the particle when (*a*) $t = 0$, (*b*) $t = 1$ s.

12.22 Slider C has a mass of 200 g and may move in a slot cut in arm AB, which rotates at the constant rate $\dot{\theta}_0 = 12$ rad/s in a horizontal plane. The slider is attached to a spring of constant $k = 36$ N/m, which is unstretched when $r = 0$. Knowing that the slider passes through the position $r = 400$ mm with a radial velocity $v_r = +1.8$ m/s, determine at that instant (*a*) the radial and transverse components of its acceleration, (*b*) its acceleration relative to arm AB, (*c*) the horizontal force exerted on the slider by arm AB.

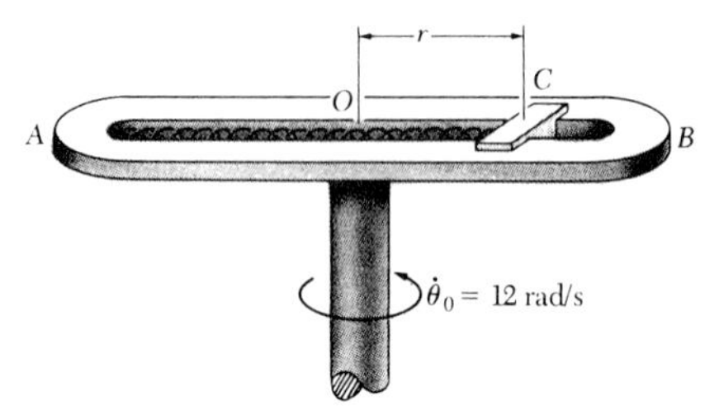

Fig. P12.22 and P12.23

12.23 Slider C has a mass of 200 g and may move in a slot cut in arm AB, which rotates at the constant rate $\dot{\theta}_0 = 12$ rad/s in a horizontal plane. The slider is attached to a spring of constant $k = 36$ N/m, which is unstretched when $r = 0$. Knowing that the slider is released with no radial velocity in the position $r = 500$ mm and neglecting friction, determine for the position $r = 300$ mm (*a*) the radial and transverse components of the velocity of the slider, (*b*) the radial and transverse components of its acceleration, (*c*) the horizontal force exerted on the slider by arm AB.

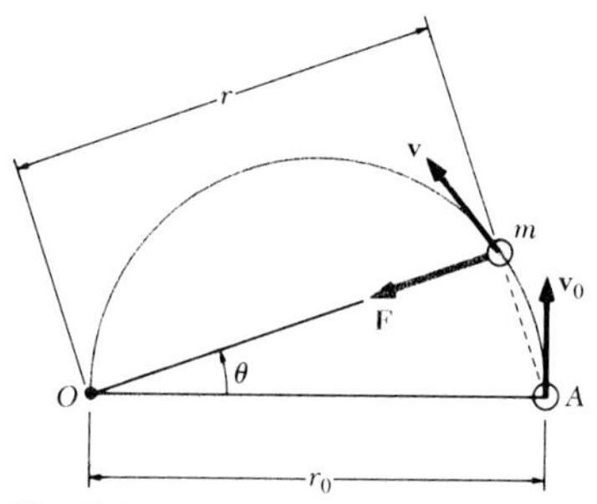

Fig. P12.24

12.24 A particle of mass m is projected from point A with an initial velocity $\mathbf{v}_0$ perpendicular to the line OA and moves under a central force $\mathbf{F}$ along a semicircular path of diameter OA. Observing that $r = r_0 \cos\theta$ and using Eq. (12.27), show that the speed v of the particle is inversely proportional to the square of the distance r from the particle to the center of force O.

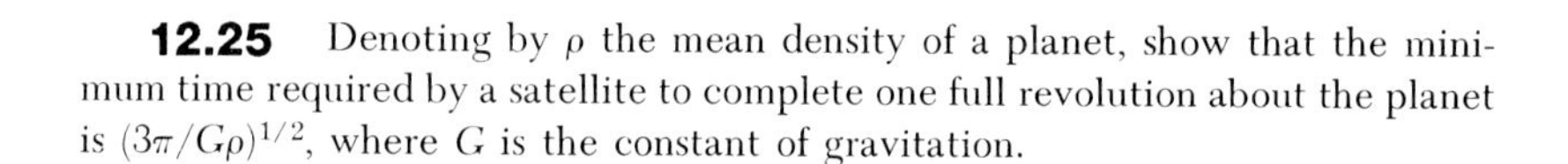
12.25 Denoting by ρ the mean density of a planet, show that the minimum time required by a satellite to complete one full revolution about the planet is $(3\pi/G\rho)^{1/2}$, where G is the constant of gravitation.

12.26 The periodic times of two of the planet Jupiter's satellites, Io and Callisto, have been observed to be, respectively, 1 day 18 h 28 min and 16 days 16 h 32 min. Knowing that the radius of Callisto's orbit is 1.884×10^6 km, determine (*a*) the mass of the planet Jupiter, (*b*) the radius of Io's orbit. (The periodic time of a satellite is the time it requires to complete one full revolution about the planet.)

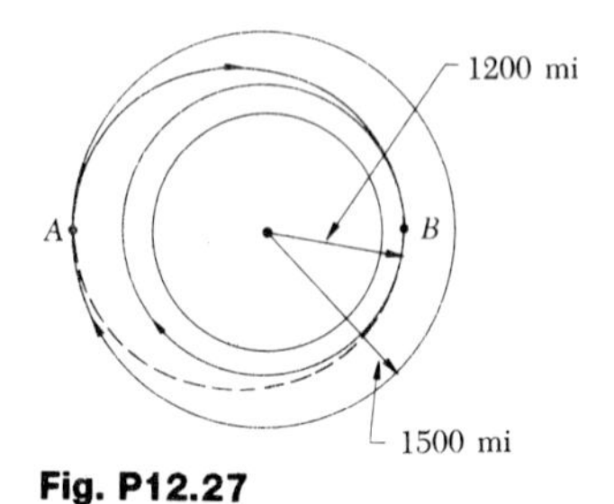

Fig. P12.27

12.27 An Apollo spacecraft describes a circular orbit with a 1500-mi radius around the moon with a velocity of 3190 mi/h. In order to transfer it to a smaller circular orbit with a 1200-mi radius, the spacecraft is first placed on an elliptic path AB by reducing its velocity to 3000 mi/h as it passes through A. Determine (*a*) the velocity of the spacecraft as it approaches B on the elliptic path, (*b*) the value to which its velocity must be reduced at B to insert it into the smaller circular orbit.

12.28 A space probe is to be placed in a circular orbit of 8000-km radius about the planet Venus in a specified plane. As the probe reaches A, the point of its original trajectory closest to Venus, it is inserted in a first elliptic transfer orbit by reducing its speed by Δv_A. This orbit brings it to point B with a much reduced velocity. There the probe is inserted in a second transfer orbit located in the specified plane by changing the direction of its velocity and further reducing its speed by Δv_B. Finally, as the probe reaches point C, it is inserted in the desired circular orbit by reducing its speed by Δv_C. Knowing that the mass of Venus is 0.82 times the mass of the earth, that $r_A = 12 \times 10^3$ km and $r_B = 96 \times 10^3$ km, that the probe reaches A with a velocity of 7400 m/s and B with a velocity of 869 m/s, and that its speed is further reduced by 146 m/s at B, determine by how much the velocity of the probe should be reduced (*a*) at A, (*b*) at C.

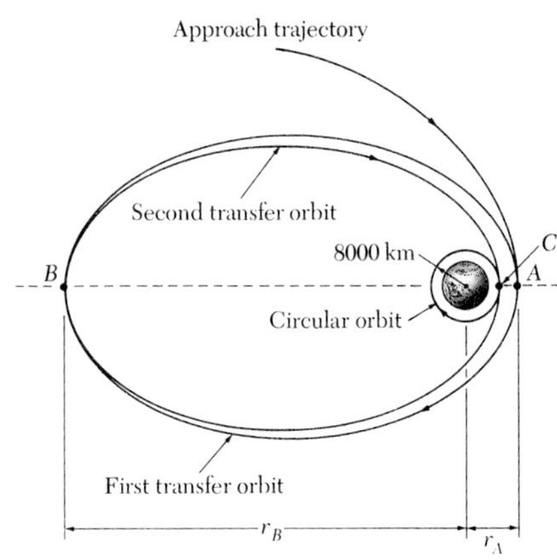

Fig. P12.28

12.29 For the space probe of Prob. 12.28, it is known that $r_A = 10 \times 10^3$ km and $r_B = 100 \times 10^3$ km, that the probe reaches A with a velocity of 8100 m/s, and that its speed is reduced by 400 m/s at A and by 2300 m/s at C. Determine (*a*) the velocity of the probe as it reaches B, (*b*) by how much its velocity should be reduced at B.

12.30 A space tug is used to place communication satellites into a geosynchronous orbit at an altitude of 22,230 mi above the surface of the earth. The tug initially describes a circular orbit at an altitude of 220 mi and is inserted in an elliptic transfer orbit by firing its engine as it passes through A, thus increasing its velocity by 7910 ft/s. By how much should its velocity be increased as it reaches B to insert it in the geosynchronous orbit?

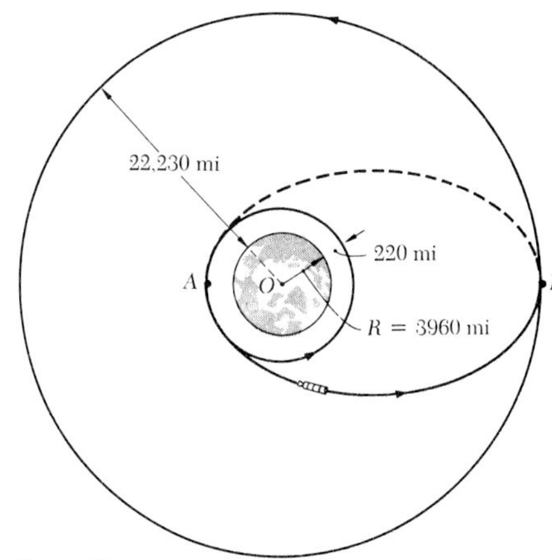

Fig. P12.30

12.31 A 5-lb ball is mounted on a horizontal rod which is free to rotate about a vertical shaft. In the position shown, the speed of the ball is $v_1 = 24$ in./s and the ball is held by a cord attached to the shaft. The cord is suddenly cut and the ball moves to position B as the rod rotates. Neglecting the mass of the rod, determine (*a*) the radial and transverse components of the acceleration of the ball immediately after the cord has been cut, (*b*) the acceleration of the ball relative to the rod at the same instant, (*c*) the speed of the ball after it has reached the stop B.

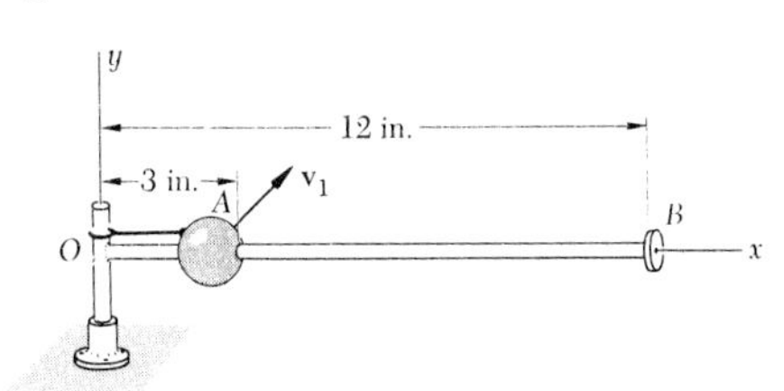

Fig. P12.31

12.32 A 4-oz ball slides on a smooth horizontal table at the end of a string which passes through a small hole in the table at O. When the length of string above the table is $r_1 = 18$ in., the speed of the ball is $v_1 = 5$ ft/s. Knowing that the breaking strength of the string is 4.00 lb, determine (*a*) the smallest distance r_2 which can be achieved by slowly drawing the string through the hole, (*b*) the corresponding speed v_2.

Fig. P12.32

12.33 A 250-g collar may slide on a horizontal rod which is free to rotate about a vertical shaft. The collar is initially held at A by a cord attached to the shaft and compresses a spring of constant 6 N/m, which is undeformed when the collar is located 500 mm from the shaft. As the rod rotates at the rate $\dot{\theta}_0 = 16$ rad/s, the cord is cut and the collar moves out along the rod. Neglecting friction and the mass of the rod, determine for the position B of the collar (*a*) the transverse component of the velocity of the collar. (*b*) the radial and transverse components of its acceleration, (*c*) the acceleration of the collar relative to the rod.

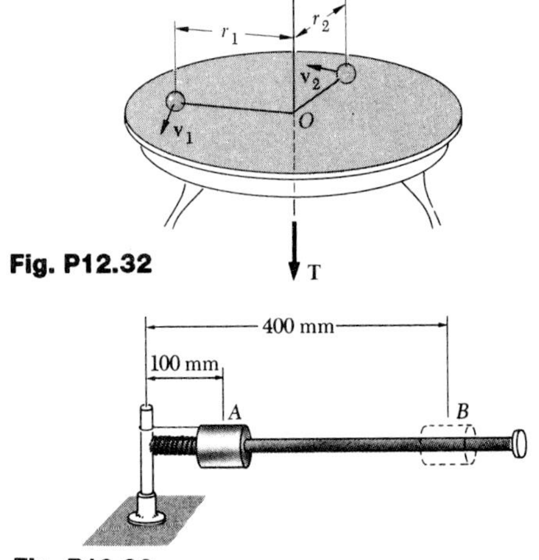

Fig. P12.33

12.34 In Prob. 12.33 determine for the position B of the collar, (*a*) the radial component of the velocity of the collar, (*b*) the value of $\ddot{\theta}$.

12.35 Two solid steel spheres, each of radius 100 mm, are placed so that their surfaces are in contact. (*a*) Determine the force of gravitational attraction between the spheres, knowing that the density of steel is 7850 kg/m³. (*b*) If the spheres are moved 2 mm apart and released with zero velocity, determine the approximate time required for their gravitational attraction to bring them back into contact. (*Hint.* Assume that the gravitational forces remain constant.)

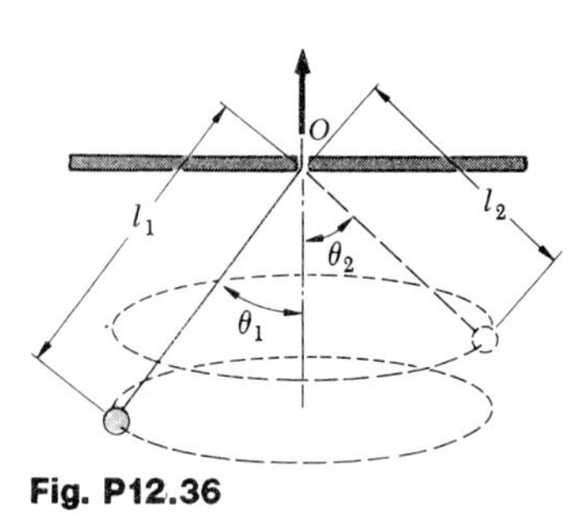

Fig. P12.36

12.36 A small ball swings in a horizontal circle at the end of a cord of length l_1, which forms an angle θ_1 with the vertical. The cord is then slowly drawn through the support at O until the length of the free end is l_2. (*a*) Derive a relation between l_1, l_2, θ_1, and θ_2. (*b*) If the ball is set in motion so that initially $l_1 = 40$ in. and $\theta_1 = 30°$, determine the angle θ_2 when $l_2 = 30$ in.

SECTIONS 12.11 to 12.13

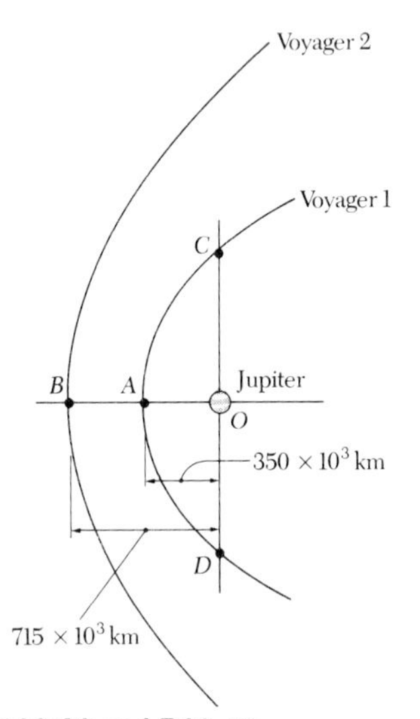

Fig. P12.39 and P12.40

12.37 A particle of mass m describes the hyperbolic spiral $r = b/\theta$ under a central force **F** directed toward the center of force O. Using Eq. (12.37), show that **F** is inversely proportional to the cube of the distance r from the particle to O.

12.38 A particle of mass m describes the logarithmic spiral $r = r_0 e^{b\theta}$ under a central force **F** directed toward the center of force O. Using Eq. (12.37), show that **F** is inversely proportional to the cube of the distance r from the particle to O.

12.39 It was observed that as the spacecraft Voyager 1 reached the point on its trajectory closest to the planet Jupiter, it was at a distance of 350×10^3 km from the center of the planet and had a velocity of 26.9 km/s. Determine the mass of Jupiter, assuming that the trajectory of the spacecraft was parabolic.

12.40 It was observed that as the spacecraft Voyager 2 reached the point on its trajectory closest to the planet Jupiter, it was at a distance of 715×10^3 km from the center of the planet. Assuming the trajectory of the spacecraft to be parabolic and using the data given in Prob. 12.39 for Voyager 1, determine the maximum velocity of Voyager 2 on its approach to Jupiter.

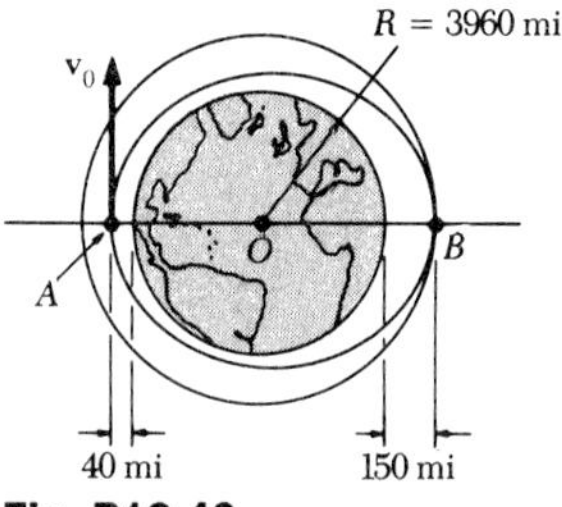

Fig. P12.43

12.41 Solve Prob. 12.40, assuming that the eccentricity of the trajectory of the spacecraft Voyager 2 was $\varepsilon = 1.20$.

12.42 It was observed that as the spacecraft Voyager 1 reached the point of its trajectory closest to the planet Saturn, it was at a distance of 115×10^3 mi from the center of the planet and had a velocity of 68.8×10^3 ft/s. Knowing that Tethys, one of Saturn's satellites, describes a circular orbit of radius 183×10^3 mi at a speed of 37.2×10^3 ft/s, determine the eccentricity of the trajectory of Voyager 1 on its approach to Saturn.

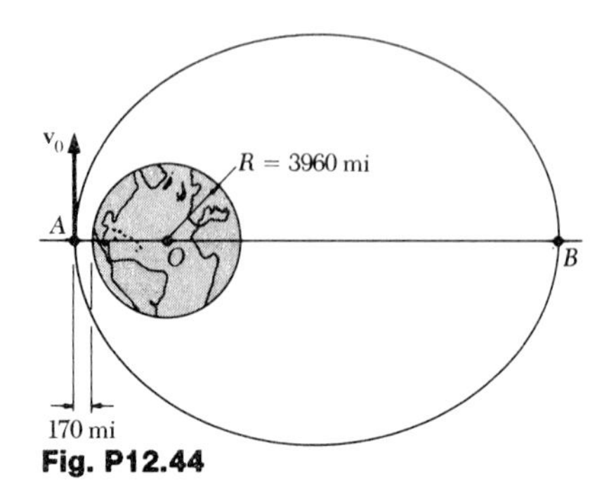

Fig. P12.44

12.43 At engine burnout on its second mission, the space shuttle Columbia had reached point A at an altitude of 40 mi and had a horizontal velocity $\mathbf{v}_0$. Knowing that its first orbit was elliptic and that the shuttle was transferred to a circular orbit as it passed through point B at an altitude of 150 mi above the surface of the earth, determine (*a*) the speed v_0 of the shuttle at engine burnout, (*b*) the increase in speed required at B to insert the shuttle on the circular orbit.

12.44 At engine burnout an Explorer satellite was 170 mi above the surface of the earth and had a horizontal velocity $\mathbf{v}_0$ of magnitude 32.6×10^3 ft/s. Determine (*a*) the highest altitude reached by the satellite, (*b*) its velocity at its apogee B.

12.45 A space probe is to be placed in a circular orbit of 8000-km radius about the planet Venus in a specified plane. As the probe reaches A, the point of its original trajectory closest to Venus, it is inserted in a first elliptic transfer orbit by reducing its speed by Δv_A. This orbit brings it to point B with a much reduced velocity. There the probe is inserted in a second transfer orbit located in the specified plane by changing the direction of its velocity and further reducing its speed by Δv_B. Finally, as the probe reaches point C, it is inserted in the desired circular orbit by reducing its speed by Δv_C. Knowing that the mass of Venus is 0.82 times the mass of the earth, that $r_A = 12.5 \times 10^3$ km and $r_B = 100 \times 10^3$ km, and that the probe approaches A on a parabolic trajectory, determine by how much the velocity of the probe should be reduced (*a*) at A, (*b*) at B, (*c*) at C.

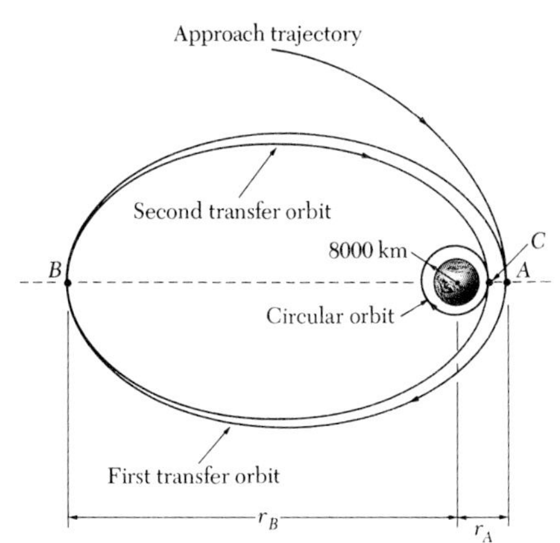

Fig. P12.45

12.46 For the space probe of Prob. 12.45, it is known that $r_A = 12.5 \times 10^3$ km and that the velocity of the probe is reduced to 6890 m/s as it passes through A. Determine (*a*) the distance from the center of Venus to point B, (*b*) the amounts by which the velocity of the probe should be reduced at B and C, respectively.

12.47 Determine the periodic time of the Explorer satellite of Prob. 12.44.

12.48 Halley's comet travels in an elongated elliptic orbit for which the minimum distance from the sun is approximately $\frac{1}{2}r_E$, where $r_E = 92.9 \times 10^6$ mi is the mean distance from the sun to the earth. Knowing that the periodic time of Halley's comet is about 76 years, determine the maximum distance from the sun reached by the comet.

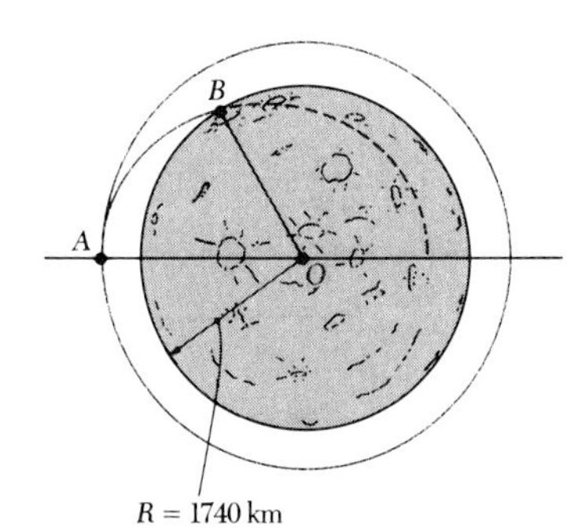

Fig. P12.49

12.49 Upon completion of their moon-exploration mission, the two astronauts forming the crew of an Apollo lunar excursion module (LEM) would rejoin the command module which had remained in a circular orbit around the moon. Before their return to earth, the astronauts would position their craft so that the LEM faced to the rear. As the command module passed through A, the LEM would be cast adrift and crash on the moon's surface at point B. Knowing that the command module was orbiting the moon at an altitude of 150 km and that the angle AOB was 60°, determine the velocity of the LEM relative to the command module as it was cast adrift. (*Hint.* Point A is the apogee of the elliptic crash trajectory. It is also recalled that the mass of the moon is 0.01230 times the mass of the earth.)

12.50 In Prob. 12.49, determine the angle AOB defining the point of impact of the LEM on the surface of the moon, assuming that the command module was orbiting the moon at an altitude of 120 km and that the LEM was cast adrift with a velocity of 40 m/s relative to the command module.

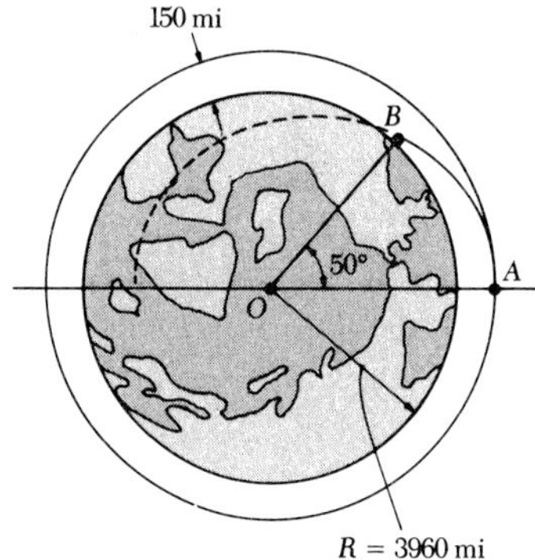

Fig. P12.51

12.51 A space shuttle is describing a circular orbit at an altitude of 150 mi above the surface of the earth. As it passes through A it fires its engine for a short interval of time to reduce its speed by 4 percent and begin its descent toward the earth. Determine the altitude of the shuttle at point B, knowing that the angle AOB is equal to 50°. (*Hint.* Point A is the apogee of the elliptic descent trajectory.)

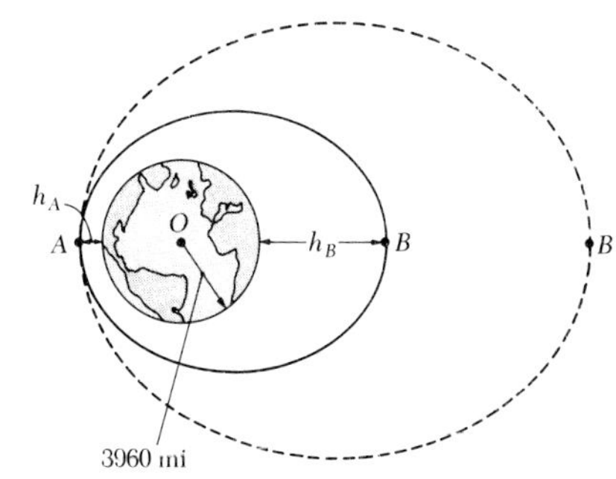

Fig. P12.52

12.52 A spacecraft is describing an elliptic orbit of minimum altitude $h_A = 1040$ mi and maximum altitude $h_B = 6040$ mi above the surface of the earth. (*a*) Determine the speed of the spacecraft at A. (*b*) If its engine is fired as the spacecraft passes through point A and its speed is increased by 10 percent, determine the maximum altitude reached by the spacecraft on its new orbit.

CHAPTER 12 COMPUTER PROBLEMS

12.C1 An airplane has a mass of 30 Mg and its engines develop a total thrust of 50 kN during take-off. The drug **D** exerted on the plane has a magnitude $D = 2.50v^2$, where v is expressed in meters per seconds and D in Newtons. The airplane starts from rest at the end of the runway and becomes airborne at a speed of 270 km/h. Derive expressions for the position and velocity of the airplane as functions of time. Plot the position and velocity as functions of time, and the velocity as a function of position as the airplane moves down the runway.

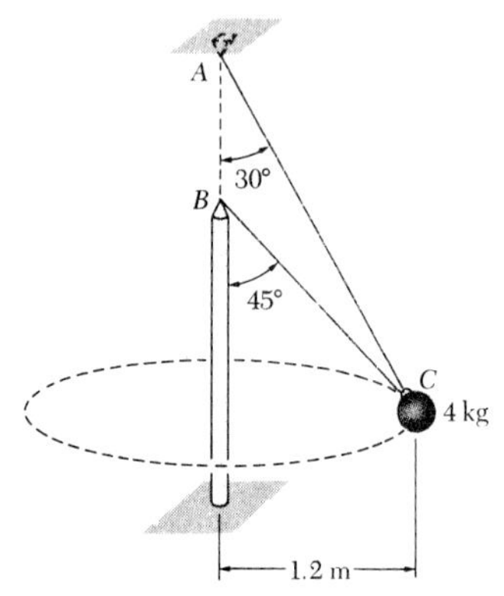

Fig. P12.C2

12.C2 Two wires AC and BC are tied at C to a sphere which revolves at the constant speed v in the horizontal circle shown. Determine and plot the tension in the wires as a function of v. Indicate on the plots the range of values of v for which both wires remain taut.

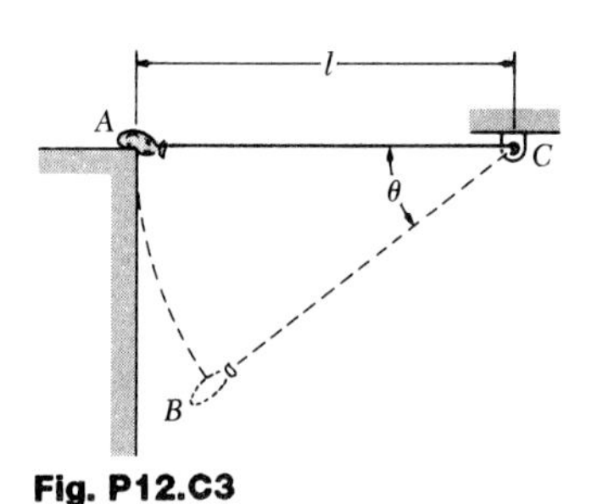

Fig. P12.C3

12.C3 A 5 kg bag is gently pushed off the top of a wall at A and swings in a vertical plane at the end of a rope of length l = 1.5 m. Determine and plot the velocity of the bag and the tension in the rope for $\theta = 0$ to $\theta = 90°$.

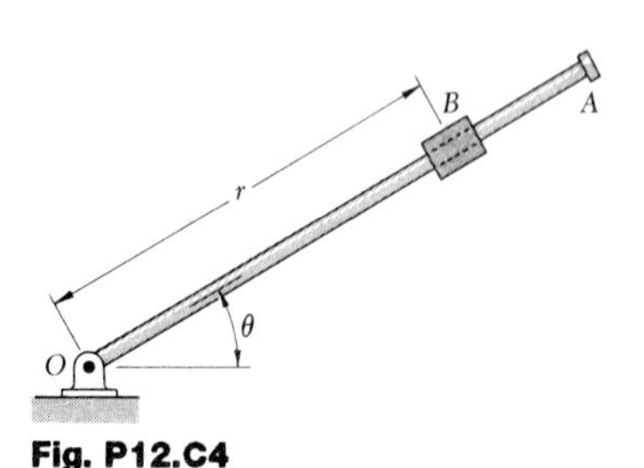

Fig. P12.C4

12.C4 A two-dimensional motion of particle B is defined by the relations $r = 25t^3 - 50t^2$ and $\theta = t^3 - 4t$, where r is expressed in millimeters, t in seconds, and θ in radians. The particle has a mass of 2 kg and moves in a horizontal plane. Determine and plot the radial and transverse components of the force acting on the particle for $t = 0$ to $t = 2.5$ s.

CHAPTER 13
KINEMATICS OF PARTICLES: ENERGY AND MOMENTUM METHODS

SECTIONS 13.1 to 13.5

13.1 A 24-lb package is placed with no initial velocity at the top of a chute. Knowing that the coefficient of kinetic friction between the package and the chute is 0.25, determine (*a*) how far the package will slide on the horizontal portion of the chute, (*b*) the maximum velocity reached by the package, (*c*) the amount of energy dissipated due to friction between *A* and *B*.

13.2 A 1050-kg trailer is hitched to a 1200-kg car. The car and trailer are traveling at 90 km/h when the driver applies the brakes on both the car and the trailer. Knowing that the braking forces exerted on the car and the trailer are 4500 N and 3600 N, respectively, determine (*a*) the distance traveled by the car and trailer before they come to a stop, (*b*) the horizontal component of the force exerted by the trailer hitch on the car.

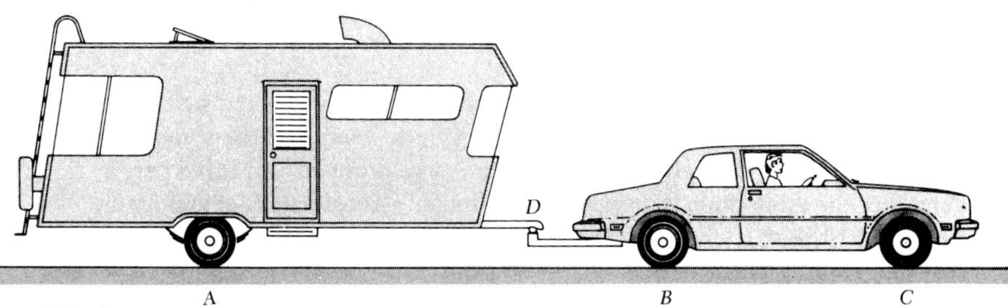

Fig. P13.2

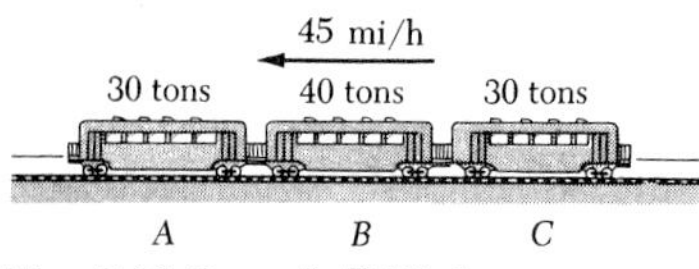

Fig. P13.3 and P13.4

13.3 The subway train shown is traveling at a speed of 45 mi/h when the brakes are fully applied on the wheels of cars A and B, causing them to slide on the track, but are not applied on the wheels of car C. Knowing that the coefficient of kinetic friction is 0.30 between the wheels and the track, determine (a) the distance required to bring the train to a stop, (b) the force in each coupling.

13.4 Solve Prob. 13.3, assuming that the brakes are applied only on the wheels of car A.

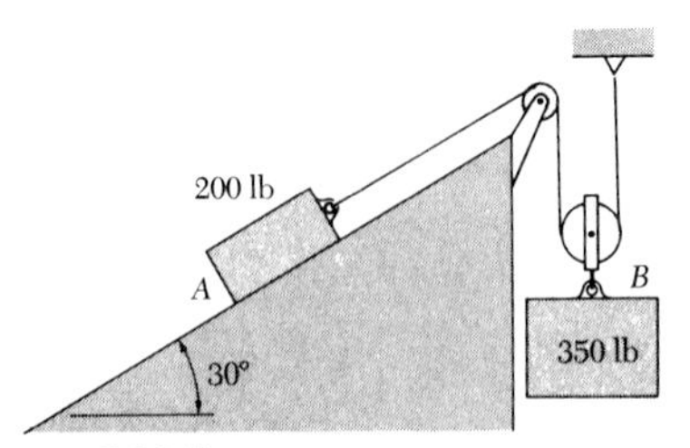

Fig. P13.5

13.5 The two blocks shown start from rest and it is observed that the velocity of block A is 6.75 ft/s after it has moved through 5 ft. Neglecting the masses of the pulleys and the friction in the pulleys, determine (a) the amount of energy dissipated due to friction between block A and the incline, (b) the coefficient of kinetic friction between block A and the incline.

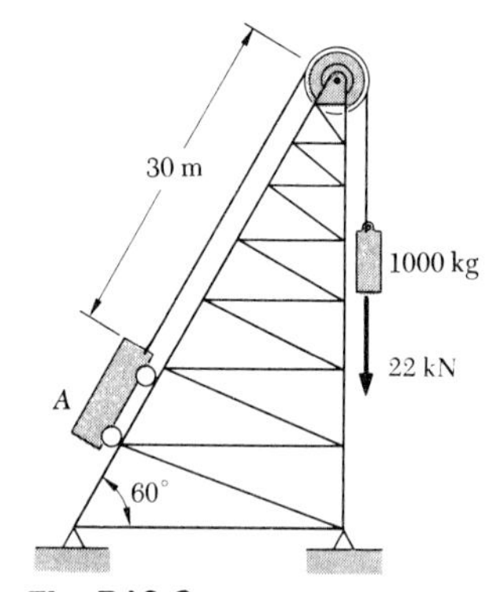

Fig. P13.6

13.6 The total mass of loading car A and its load is 3500 kg. The car is connected to a 1000-kg counterweight and is at rest when a constant 22-kN force is applied as shown. (a) If the force acts through the entire motion, what is the speed of the car after it has traveled 30 m? (b) If after the car has moved through a distance x the 22-kN force is removed, the car will coast to rest. After what distance x should the force be removed if the car is to come to rest after a total movement of 30 m?

13.7 Knowing that the system shown starts from rest, determine (a) the velocity of collar A after it has moved through 320 mm, (b) the corresponding velocity of collar B, (c) the tension in the cable. Neglect the masses of the pulleys and the effect of friction.

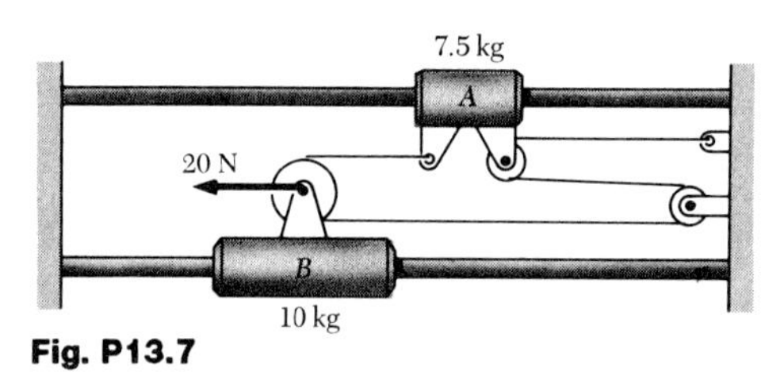

Fig. P13.7

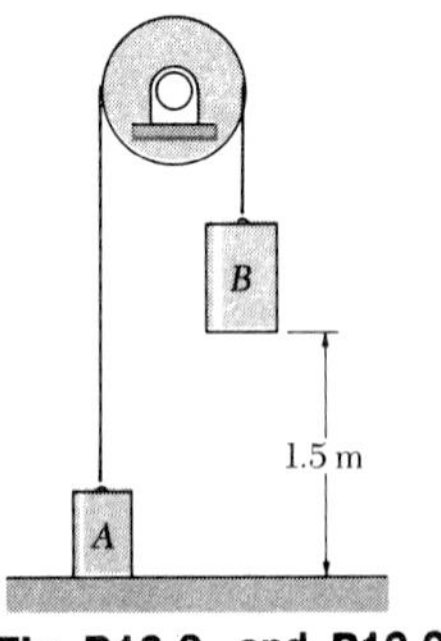

Fig. P13.8 and P13.9

13.8 Two blocks A and B, of mass 8 kg and 12 kg, respectively, hang from a cable which passes over a pulley of negligible mass. Knowing that the blocks are released from rest and that the energy dissipated by axle friction in the pulley is 10 J, determine (a) the velocity of block B as it strikes the ground, (b) the force exerted by the cable on each of the two blocks during the motion.

13.9 Two blocks A and B, of mass 12 kg and 15 kg, respectively, hang from a cable which passes over a pulley of negligible mass. The blocks are released from rest in the positions shown and block B is observed to strike the ground with a velocity of 1.6 m/s. Determine (a) the energy dissipated due to axle friction in the pulley, (b) the force exerted by the cable on each of the two blocks during the motion.

13.10 Two blocks A and D, weighing, respectively, 125 lb and 300 lb, are attached to a rope which passes over two fixed pipes B and C as shown. It is observed that when the system is released from rest, block A acquires a velocity of 8 ft/s after moving 5 ft up. Determine (a) the force exerted by the rope on each of the two blocks during the motion, (b) the coefficient of kinetic friction between the rope and the pipes, (c) the energy dissipated due to friction.

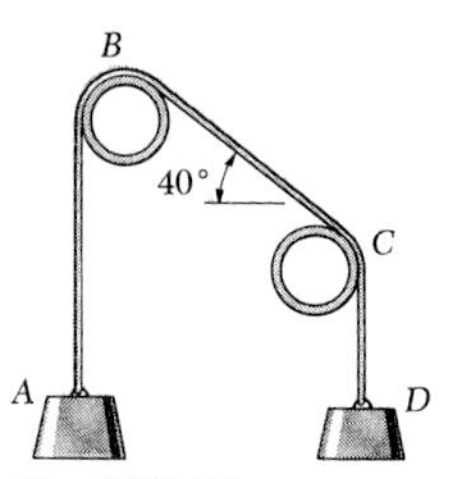

Fig. P13.10

13.11 An 8-kg plunger is released from rest in the position shown and is stopped by two nested springs; the constant of the outer spring is $k_1 = 4$ kN/m and the constant of the inner spring is $k_2 = 12$ kN/m. If the maximum deflection of the outer spring is observed to be 125 mm, determine the height h from which the plunger was released.

13.12 An 8-kg plunger is released from rest in the position shown and is stopped by two nested springs; the constant of the outer spring is $k_1 = 4$ kN/m and the constant of the inner spring is $k_2 = 12$ kN/m. If the plunger is released from the height $h = 600$ mm, determine the maximum deflection of the outer spring.

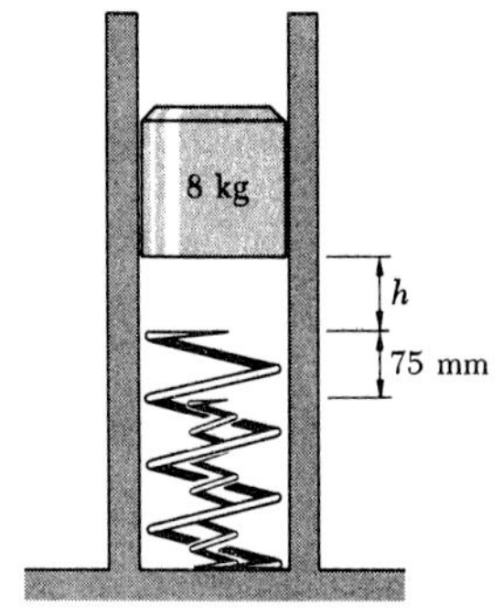

Fig. P13.11 and P13.12

13.13 A 120-mm-diameter piston having a mass of 5 kg slides without friction in a cylinder. The pressure p within the cylinder varies inversely as the volume of the cylinder and is equal to the atmospheric pressure $p_a = 101.3$ kN/m^2 when $x = 200$ mm. If the piston is moved to the left and released with no velocity when $x = 80$ mm, determine the maximum velocity reached by the piston in the ensuing motion.

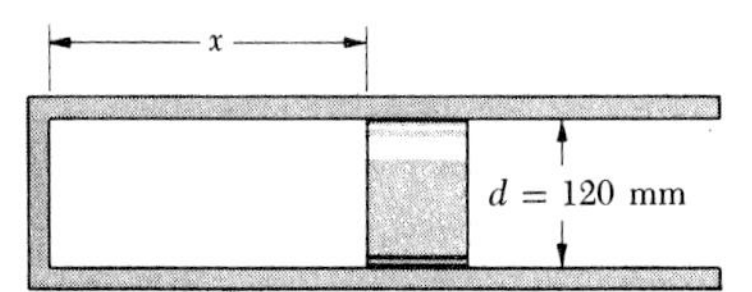

Fig. P13.13 and P13.14

13.14 In Prob. 13.13, determine the maximum value of the coordinate x after the piston has been released with no velocity in the position $x = 80$ mm.

13.15 A bullet is fired straight up from the surface of the moon with an initial velocity of 500 m/s. Determine the maximum elevation reached by the bullet, (a) assuming a uniform gravitational field with $g = 1.62$ m/s^2, (b) using Newton's law of gravitation. (Radius of moon = 1740 km.)

13.16 A rocket is fired vertically from the ground. Knowing that at burnout the rocket is 50 mi above the ground and has a velocity of 15,000 ft/s, determine the highest altitude it will reach.

13.17 Sphere C and block A are both moving to the left with a velocity $\mathbf{v}_0$ when the block is suddenly stopped by the wall. Determine the smallest velocity $\mathbf{v}_0$ for which the sphere C will swing in a full circle about the pivot B (a) if BC is a slender rod of negligible mass, (b) if BC is a cord.

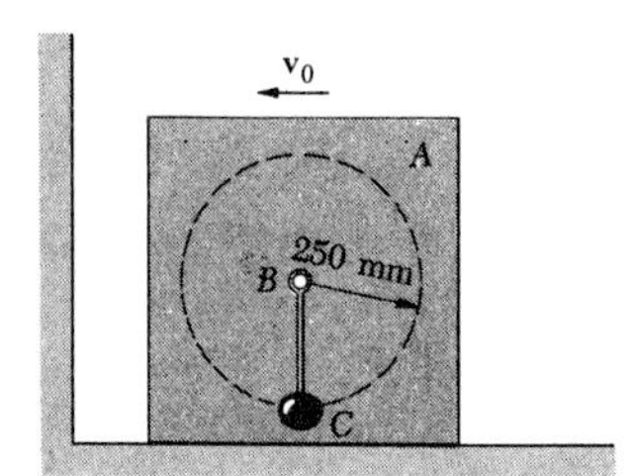

Fig. P13.17

13.18 A bag is gently pushed off the top of a wall at A and swings in a vertical plane at the end of a rope of length l. Determine the angle θ for which the rope will break, knowing that it can withstand a maximum tension 50 percent larger than the weight of the bag.

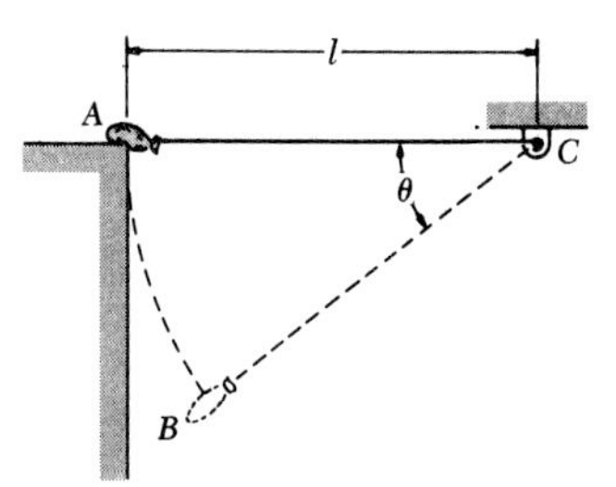

Fig. P13.18

13.19 A 15-g bullet leaves a fixed rifle barrel 2 ms after being fired. Knowing that the muzzle velocity is 800 m/s and neglecting friction, determine the average power developed by the rifle.

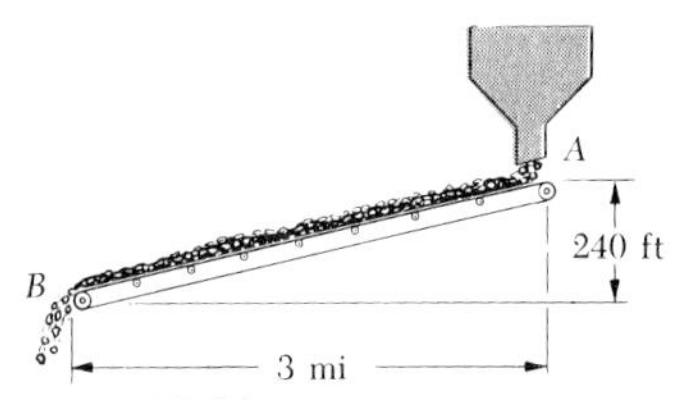

Fig. P13.21

13.20 A 75-kg man runs up a 6-m-high flight of stairs in 8 s. (*a*) What is the average power developed by the man? (*b*) If a 60-kg woman can develop 80 percent as much power, how long will it take her to run up a 4-m-high flight of stairs?

13.21 Crushed stone is being moved from a quarry at *A* to a loading dock at *B* at the rate of 500 tons/h. An electric generator is attached to the system in order to maintain a constant belt speed. Knowing that the efficiency of the belt-generator system is 0.70, determine the average power in kilowatts developed by the generator if the belt speed is (*a*) 6 ft/s, (*b*) 10 ft/s.

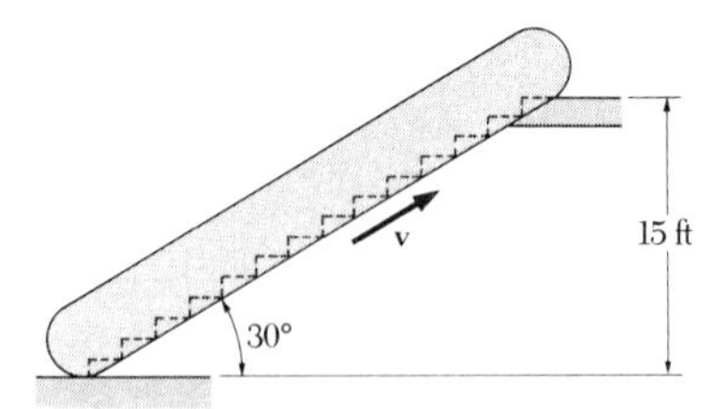

Fig. P13.22

13.22 The escalator shown is designed to transport 6000 persons per hour at a constant speed of 1.5 ft/s. Assuming an average weight of 150 lb per person, determine (*a*) the average power required, (*b*) the required capacity of the motor if the mechanical efficiency is 85 percent and if a 300 percent overload is to be allowed.

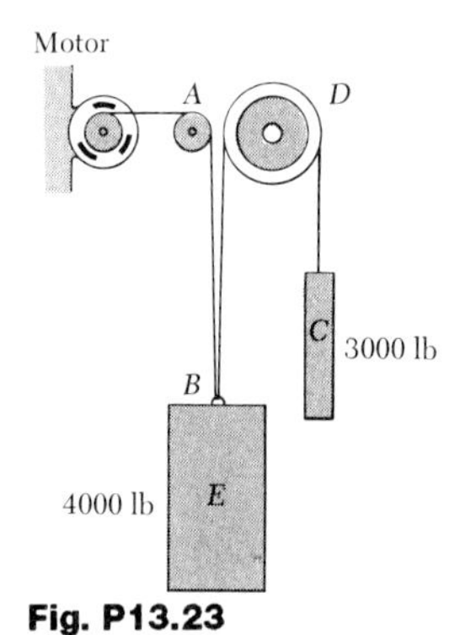

Fig. P13.23

13.23 Elevator *E* weighs 4000 lb when fully loaded and is connected to a 3000-lb counterweight *C*. Determine the power delivered by the electric motor when the elevator (*a*) is moving up at a constant speed of 20 ft/s, (*b*) has an instantaneous velocity of 20 ft/s and an acceleration of 3 ft/s^2, both directed upward.

13.24 The fluid transmission of a 12-Mg truck permits the engine to deliver an essentially constant power of 40 kW to the driving wheels. Determine the time required and the distance traveled as the speed of the truck is increased (*a*) from 24 km/h to 48 km/h, (*b*) from 48 km/h to 72 km/h.

SECTIONS 13.6 to 13.9

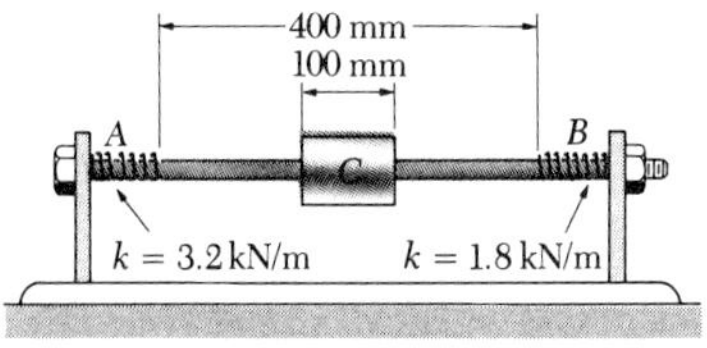

Fig. P13.25

13.25 A collar *C* of mass *m* slides without friction on a horizontal rod between springs *A* and *B*. If the collar is pushed to the left until spring *A* is compressed 60 mm and released, determine the distance through which the collar will travel and the maximum velocity it will reach (*a*) if $m = 0.5$ kg, (*b*) if $m = 2$ kg.

13.26 A 0.8-kg collar *C* may slide without friction along a horizontal rod. It is attached to a spring of constant $k = 300$ N/m and 500-mm undeformed length. Knowing that the collar is released from rest in the position shown, determine the maximum velocity it will reach in the ensuing motion.

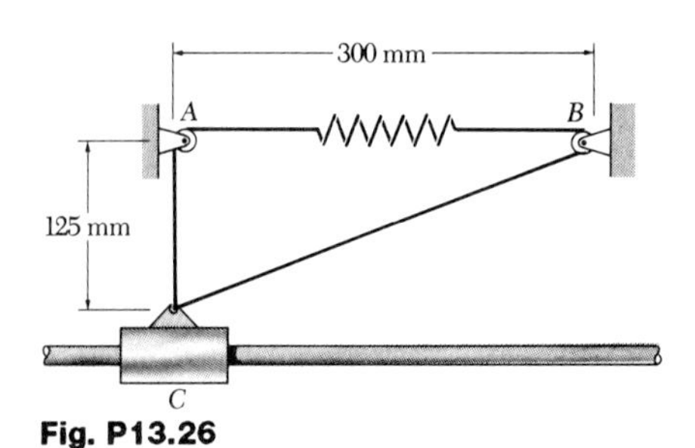

Fig. P13.26

13.27 An elastic cord is stretched between two points A and B, located 12 in. apart in the same horizontal plane. When stretched directly between A and B, the tension in the cord is 5 lb. The cord is then stretched as shown until its midpoint C has moved through 4.5 in. to C'; a force of 30 lb is required to hold the cord at C'. A 2-oz pellet is placed at C', and the cord is released. Determine the speed of the pellet as it passes through C.

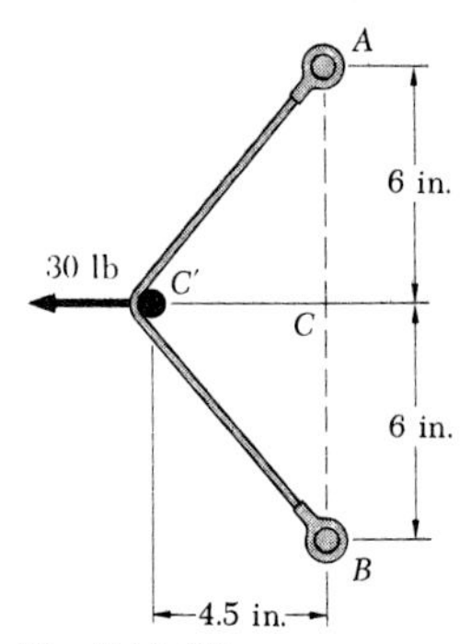

Fig. P13.27

13.28 A 600-g collar C may slide along a horizontal, semicircular rod ABD. The spring CE has an undeformed length of 250 mm and a spring constant of 135 N/m. Knowing that the collar is released from rest at A and neglecting friction, determine the speed of the collar (*a*) at B, (*b*) at D.

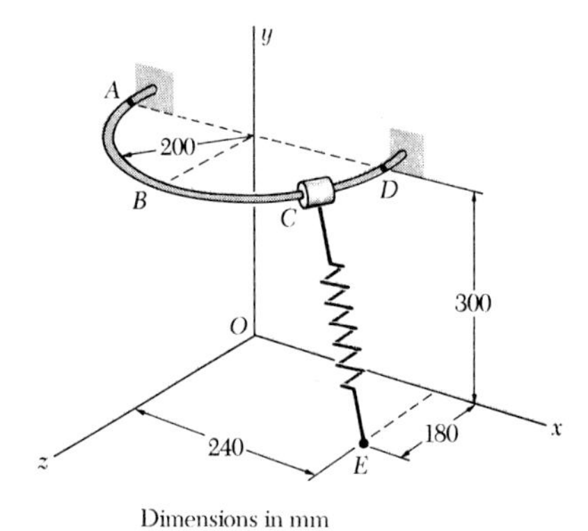

Fig. P13.28

13.29 A 30-Mg railroad car starts from rest and coasts down a 1 percent incline for a distance of 20 m. It is stopped by a bumper having a spring constant of 2400 kN/m. (*a*) What is the speed of the car at the bottom of the incline? (*b*) How many millimeters will the spring be compressed?

13.30 A toy spring gun is used to shoot 1-oz bullets vertically upward. The undeformed length of the spring is 6 in.; it is compressed to a length of 1 in. when the gun is ready to be shot and expands to a length of 3 in. as the bullet leaves the gun. A force of 10 lb is required to maintain the spring in firing position when the length of the spring is 1 in. Determine (*a*) the velocity of the bullet as it leaves the gun, (*b*) the maximum height reached by the bullet.

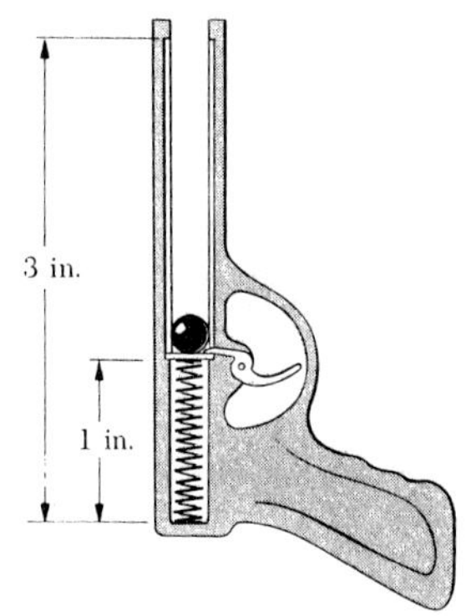

Fig. P13.30

13.31 A spring is used to stop a 150-lb package which is moving down a 20° incline. The spring has a constant $k = 150$ lb/in. and is held by cables so that it is initially compressed 4 in. Knowing that the velocity of the package is 10 ft/s when it is 30 ft from the spring and neglecting friction, determine the maximum additional deformation of the spring in bringing the package to rest.

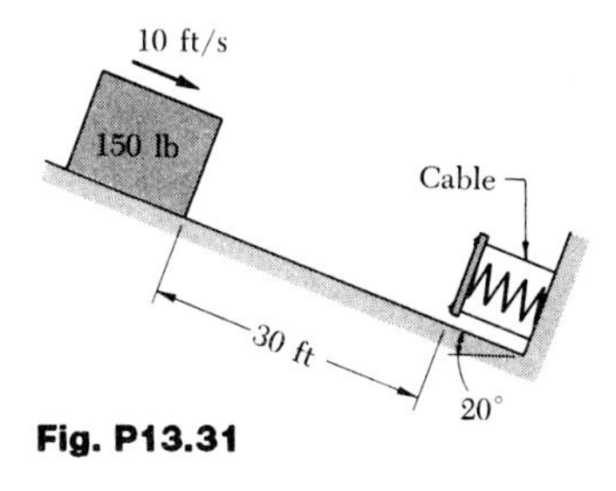

Fig. P13.31

13.32 Two springs are attached to a piece of cloth A of negligible mass, as shown. The initial tension in each spring is 500 N, and the spring constant of each spring is $k = 2$ kN/m. A 20-kg ball is released from a height h above A; the ball hits the cloth, causing it to move through a maximum distance $d = 0.9$ m. Determine the height h.

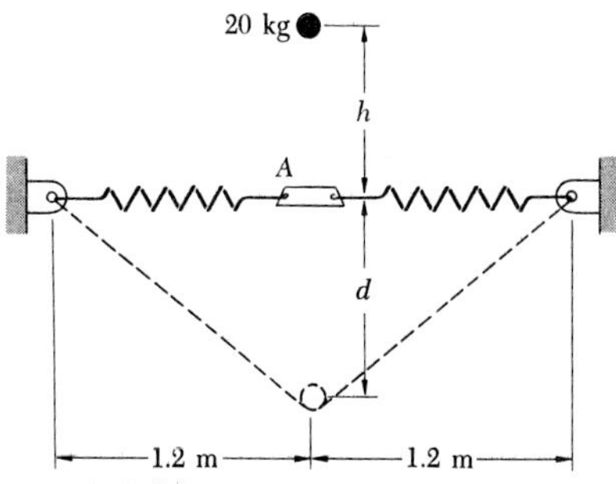

Fig. P13.32

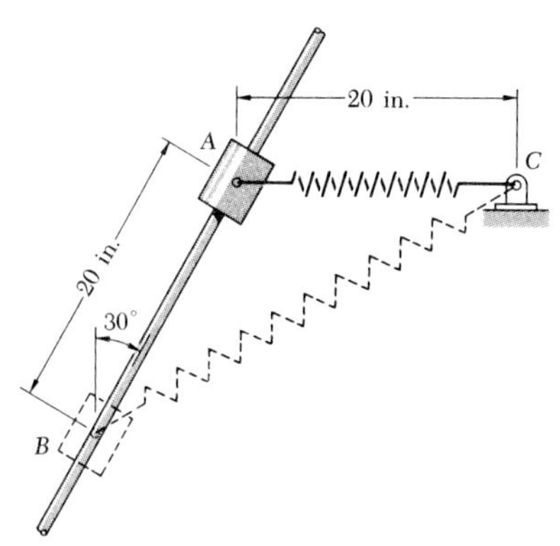

Fig. P13.33

13.33 A 12-lb collar slides without friction along a rod which forms an angle of 30° with the vertical. The spring is unstretched when the collar is at A. If the collar is released from rest at A, determine the value of the spring constant k for which the collar has zero velocity at B.

13.34 A $\frac{1}{2}$-lb pellet is released from rest at A when the spring is compressed 3 in. and travels around the loop $ABCDE$. Determine the smallest value of the spring constant for which the pellet will travel around the loop and will at all times remain in contact with the loop.

13.35 The pendulum shown is released from rest at A and swings through 90° before the cord touches the fixed peg B. Determine the smallest value of a for which the pendulum bob will describe a circle about the peg.

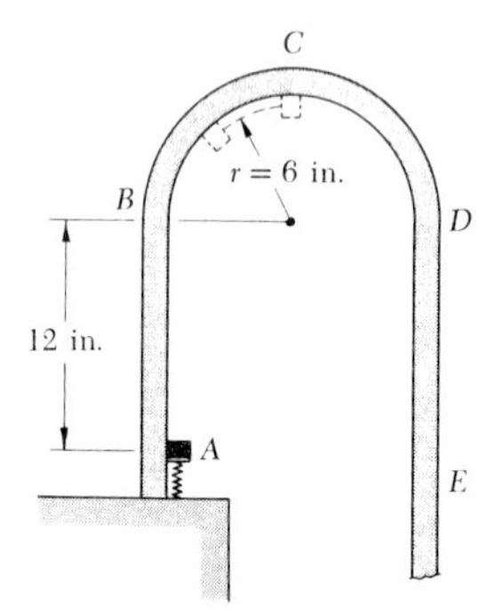

Fig. P13.34

13.36 (a) Determine the kinetic energy per unit mass which a missile must have after being fired from the surface of the earth if it is to reach an infinite distance from the earth. (b) What is the initial velocity of the missile (called *escape velocity*)? Give your answers in SI units and show that the answer to part b is independent of the firing angle.

13.37 During its fifth mission, the space shuttle Columbia ejected two communication satellites while describing a circular orbit, 185 mi above the surface of the earth. Knowing that one of these satellites weighed 8000 lb, determine (a) the additional energy required to place the satellite in a geosynchronous orbit (see Prob. 12.79) at an altitude of 22,230 mi above the surface of the earth, (b) the energy required to place it in the same orbit by launching it from the surface of the earth.

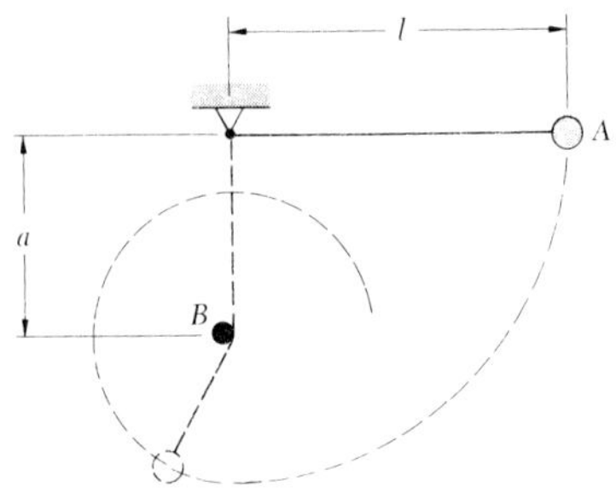

Fig. P13.35

13.38 A 200-g ball may slide on a horizontal frictionless surface and is attached to a fixed point O by means of an elastic cord of constant $k = 150$ N/m and undeformed length equal to 600 mm. The ball is placed at point A, 900 mm from O, and is given an initial velocity $\mathbf{v}_A$ in a direction perpendicular to OA. Knowing that the ball passes at a distance $d = 100$ mm from point O, determine (a) the initial speed v_A of the ball, (b) its speed v after the cord has become slack.

13.39 Knowing that the initial speed of the ball of Prob. 13.38 is $v_A = 3$ m/s, determine (a) the speed v of the ball after the cord has become slack, (b) the closest distance d that the ball will come to O.

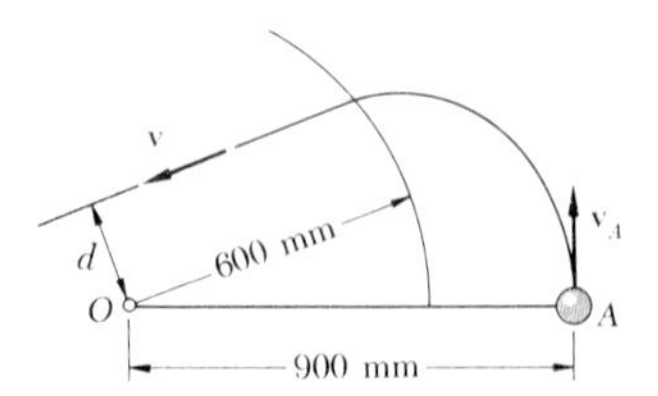

Fig. P13.38

13.40 A $\frac{1}{2}$-lb collar may slide on a horizontal rod which is free to rotate about a vertical shaft. The collar is initially held at A by a cord attached to the shaft and compresses a spring of constant 2.5 lb/ft, which is undeformed when the collar is located 9 in. from the shaft. As the rod rotates at the rate $\dot{\theta}_0 = 12$ rad/s, the cord is cut and the collar moves along the rod. Neglecting friction and the mass of the rod, and assuming that the spring is attached to the collar and to the shaft, determine the radial and transverse components of the velocity of the collar as it passes through B.

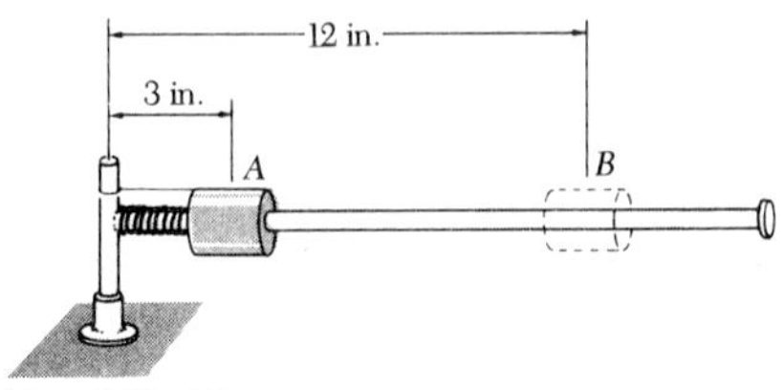

Fig. P13.40

13.41 At engine burnout, an experimental rocket has reached an altitude of 400 km and has a velocity $\mathbf{v}_0$ of magnitude 7500 m/s forming an angle of 35° with the vertical. Determine the maximum altitude reached by the rocket.

13.42 A 50,000-ton ocean liner has an initial velocity of 3 mi/h. Neglecting the frictional resistance of the water, determine the time required to bring the liner to rest by using a single tugboat which exerts a constant force of 45 kips.

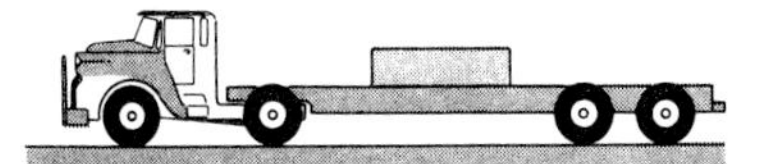

Fig. P13.44

13.43 A 1200-kg automobile is moving at a speed of 90 km/h when the brakes are fully applied, causing all four wheels to skid. Determine the time required to stop the automobile (*a*) on dry pavement ($\mu_k = 0.75$), (*b*) on an icy road ($\mu_k = 0.10$).

13.44 The coefficients of friction between the load and the flatbed trailer shown are $\mu_s = 0.50$ and $\mu_k = 0.40$. Knowing that the speed of the rig is 72 km/h, determine the shortest time in which the rig can be brought to a stop if the load is not to shift.

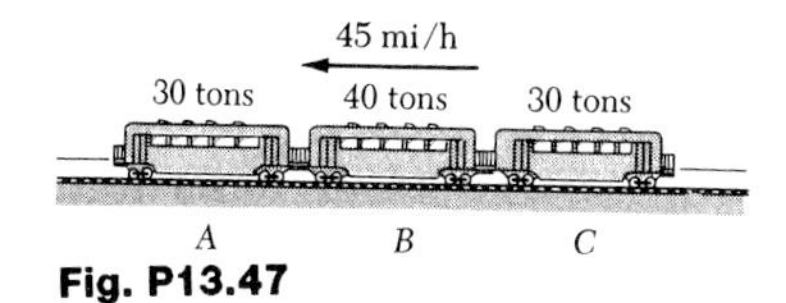

Fig. P13.47

13.45 A 10-lb particle is acted upon by the force, expressed in pounds, $\mathbf{F} = (4 - 3t)\mathbf{i} + (2 - t^2)\mathbf{j} + (2 + t)\mathbf{k}$. Knowing that at $t = 0$ the velocity of the particle is $\mathbf{v} = (75\text{ ft/s})\mathbf{i} + (50\text{ ft/s})\mathbf{j} - (125\text{ ft/s})\mathbf{k}$, determine (*a*) the time at which the velocity of the particle is parallel to the *yz* plane, (*b*) the corresponding velocity of the particle.

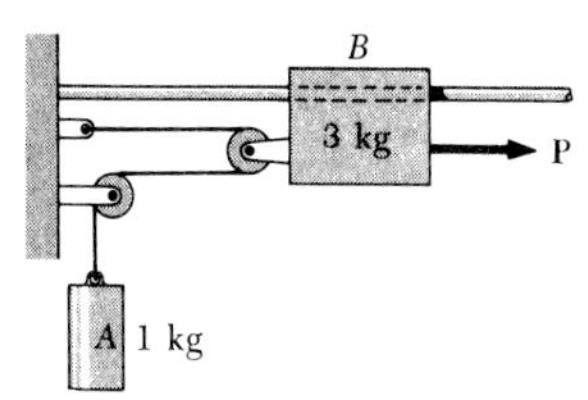

Fig. P13.48

13.46 A 5-kg particle is acted upon by the force, expressed in newtons, $\mathbf{F} = 10t\,\mathbf{i} + (15 - 6t)\mathbf{j} + 4t^3\mathbf{k}$. Knowing that the velocity of the particle at $t = 0$ is $\mathbf{v} = -(8\text{ m/s})\mathbf{i} + (5\text{ m/s})\mathbf{j} - (20\text{ m/s})\mathbf{k}$, determine the velocity of the particle at $t = 5$ s.

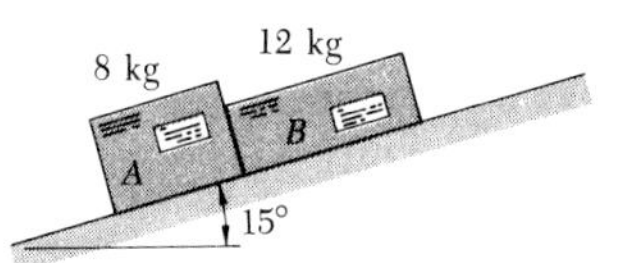

Fig. P13.49

13.47 The subway train shown is traveling at a speed of 45 mi/h when the brakes are fully applied on the wheels of cars *A* and *B*, causing them to slide on the track, but are not applied on the wheels of car *C*. Knowing that the coefficient of kinetic friction is 0.30 between the wheels and the track, determine (*a*) the time required to bring the train to a stop, (*b*) the force in each coupling.

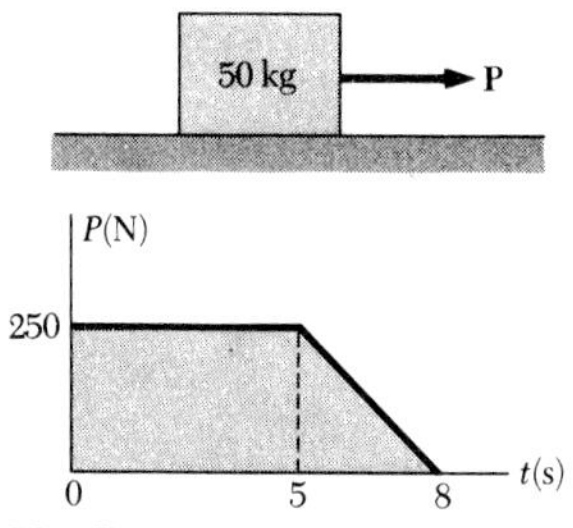

Fig. P13.50

13.48 The system shown is initially at rest. Neglecting friction, determine (*a*) the force **P** required if the velocity of collar *B* is to be 5 m/s after 2 s, (*b*) the corresponding tension in the cable.

13.49 Two packages are placed on an incline as shown. The coefficients of friction are $\mu_s = 0.25$ and $\mu_k = 0.20$ between the incline and package *A*, and $\mu_s = 0.15$ and $\mu_k = 0.12$ between the incline and package *B*. Knowing that the packages are in contact when released, determine (*a*) the velocity of each package after 3 s, (*b*) the force exerted by package *A* on package *B*.

13.50 A 50-kg block initially at rest is acted upon by a force **P** which varies as shown. Knowing that the coefficient of kinetic friction between the block and the horizontal surface is 0.20, determine the velocity of the block (*a*) at $t = 5$ s, (*b*) at $t = 8$ s.

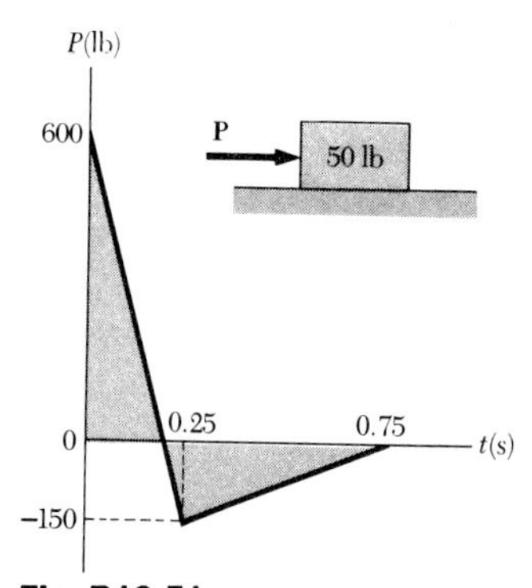

Fig. P13.51

13.51 The pressure wave produced by an explosion exerts on a 50-lb block a force **P** whose variation with time may be approximated as shown. Knowing that the block was initially at rest and neglecting the effect of friction, determine (*a*) the maximum velocity reached by the block, (*b*) the velocity of the block at $t = 0.75$ s.

13.52 A 30-g bullet is fired with a velocity of 640 m/s into a wooden block which rests against a solid vertical wall. Knowing that the bullet is brought to rest in 0.8 ms, determine the average impulsive force exerted by the bullet on the block.

13.53 A 2500-lb car moving with a velocity of 3 mi/h hits a garage wall and is brought to rest in 0.05 s. Determine the average impulsive force exerted by the wall on the car bumper.

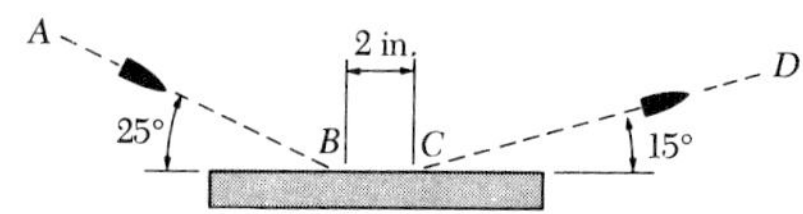

Fig. P13.54

13.54 A 1-oz steel-jacketed bullet is fired with a velocity of 2200 ft/s toward a steel plate and ricochets along the path CD with a velocity of 1800 ft/s. Knowing that the bullet leaves a 2-in. scratch on the surface of the plate and assuming that it has an average speed of 2000 ft/s while in contact with the plate, determine the magnitude and direction of the impulsive force exerted by the plate on the bullet.

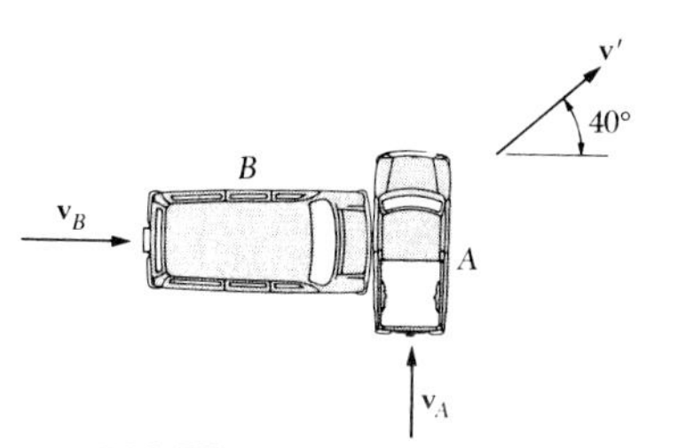

Fig. P13.55

13.55 Car A was traveling due north through an intersection when it was hit broadside by car B which was traveling due east. While both drivers admitted having ignored the four-way stop signs at the intersection, each claimed that he was traveling at the 35-mi/h speed limit and that the other was traveling much faster. Knowing that car A weighs 2000 lb, car B 3600 lb, and that inspection of the scene of the accident showed that as a result of the impact, the two cars got stuck together and skidded in a direction 40° north of east, determine (*a*) which of the two cars was actually traveling at 35 mi/h, (*b*) how fast the other car was moving.

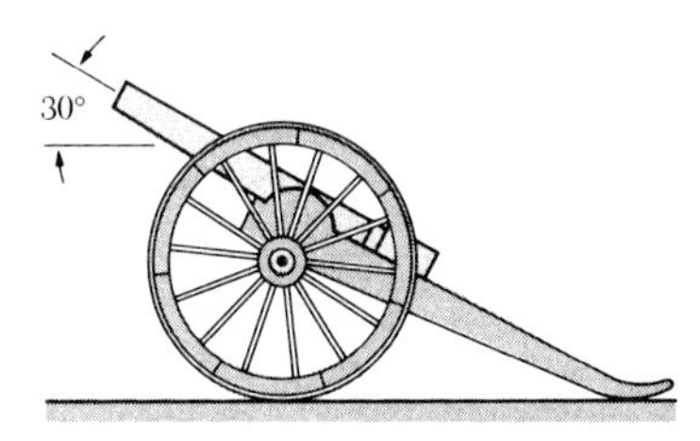

Fig. P13.56

13.56 An old 4000-lb gun fires a 20-lb shell with an initial velocity of 2000 ft/s at an angle of 30°. The gun rests on a horizontal surface and is free to move horizontally. Assuming that the barrel of the gun is rigidly attached to the frame (no recoil mechanism) and that the shell leaves the barrel 6 ms after firing, determine (*a*) the recoil velocity of the gun, (*b*) the resultant **R** of the vertical impulsive forces exerted by the ground on the gun.

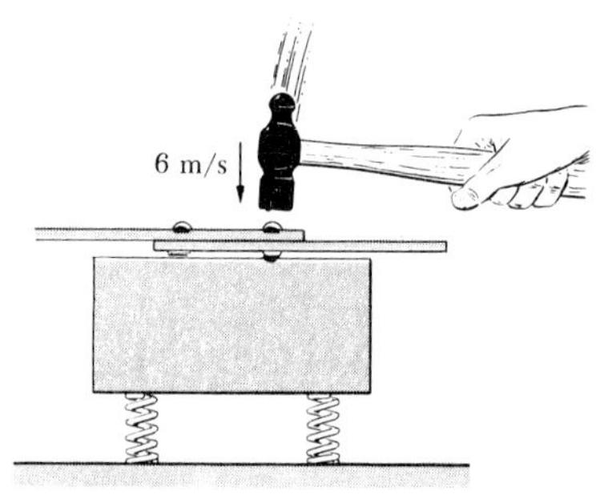

Fig. P13.57

13.57 A small rivet connecting two pieces of sheet metal is being clinched by hammering. Determine the impulse exerted on the rivet and the energy absorbed by the rivet under each blow, knowing that the head of the hammer has a mass of 800 g and that it strikes the rivet with a velocity of 6 m/s. Assume that the hammer does not rebound and that the anvil is supported by springs and (*a*) has an infinite mass (rigid support), (*b*) has a mass of 4 kg.

SECTION 13.12 to 13.15

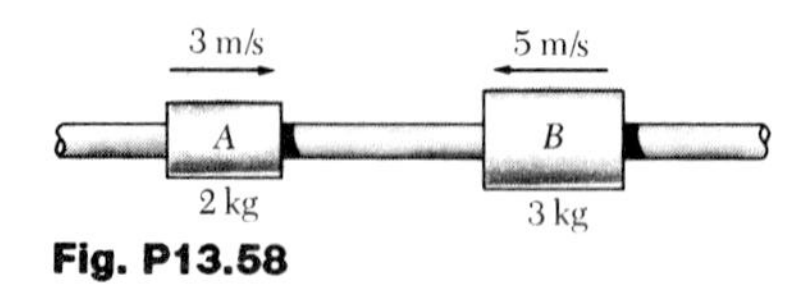

Fig. P13.58

13.58 The velocities of two collars before impact are as shown. If after impact the velocity of collar A is observed to be 5.4 m/s to the left, determine (*a*) the velocity of collar B after impact, (*b*) the coefficient of restitution between the two collars.

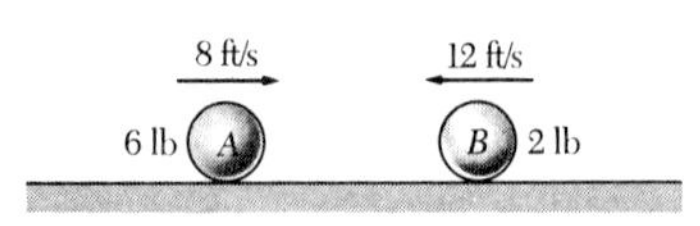

Fig. P13.59

13.59 Two small spheres A and B are made of different materials and have the masses indicated. They are moving on a frictionless, horizontal surface with the velocities shown when they hit each other. Knowing that the coefficient of restitution between the spheres is $e = 0.80$, determine (*a*) the velocity of each sphere after impact, (*b*) the energy loss due to the impact.

13.60 A 2.5-lb ball A is falling vertically with a velocity of magnitude $v_A = 8$ ft/s when it is hit as shown by a 1.5-lb ball B which has a velocity of magnitude $v_B = 5$ ft/s. Knowing that the coefficient of restitution between the two balls is $e = 0.75$ and assuming no friction, determine the velocity of each ball immediately after impact.

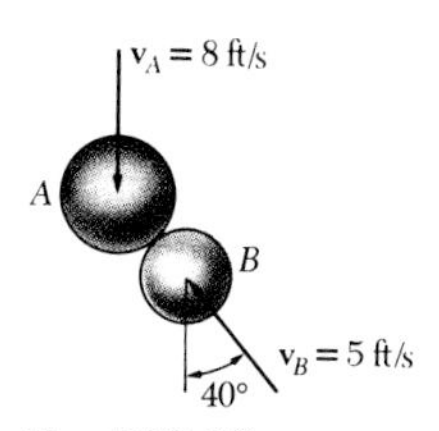

Fig. P13.60

13.61 Two identical billiard balls may move freely on a horizontal table. Ball A has a velocity $\mathbf{v}_0$ as shown and hits ball B, which is at rest, at a point C defined by $\theta = 30°$. Knowing that the coefficient of restitution between the two balls is $e = 0.90$ and assuming no friction, determine the velocity of each ball after impact.

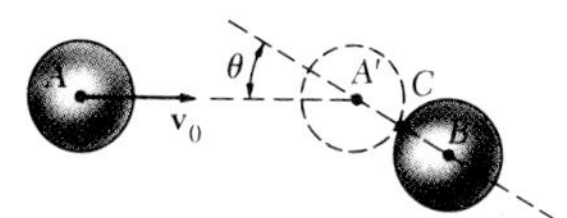

Fig. P13.61

13.62 A ball falling vertically hits a 90° corner at B with a velocity $\mathbf{v}_0$. Show that after hitting the corner again at C, it will rebound vertically with a velocity $-e\mathbf{v}_0$, where e is the coefficient of restitution between the ball and the corner. Neglect friction and assume that the short intermediate path BC is a straight line.

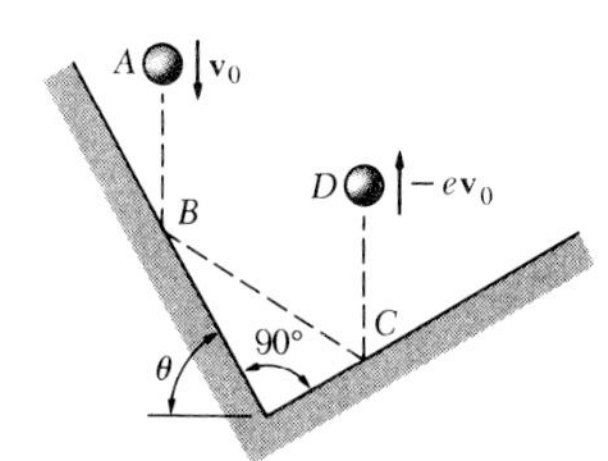

Fig. P13.62

13.63 A small ball A is dropped from a height h onto a rigid, frictionless plate at B and bounces to point C at the same elevation as B. Knowing that $\theta = 20°$ and that the coefficient of restitution between the ball and the plate is $e = 0.40$, determine the distance d.

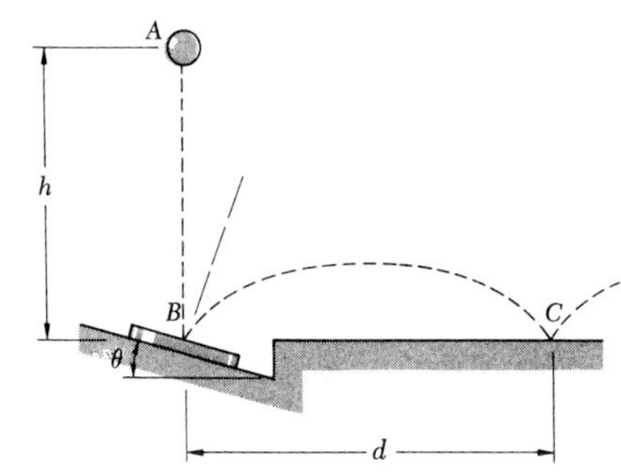

Fig. P13.63

13.64 A ball moving with the horizontal velocity $\mathbf{v}_0$ drops from A through the vertical distance $h_0 = 10$ in. to a frictionless floor. Knowing that the ball hits the floor at a distance $d_0 = 4$ in. from B and that the coefficient of restitution between the ball and the floor is $e = 0.80$, determine (*a*) the height h_1 and length d_1 of the first bounce, (*b*) the height h_2 and length d_2 of the second bounce.

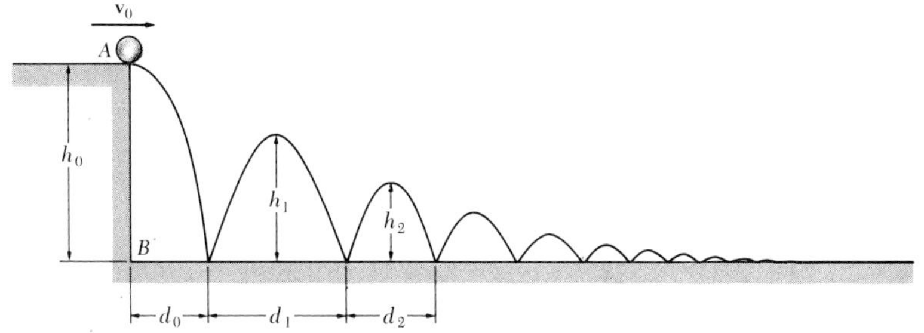

Fig. P13.64, P13.65, and P13.66

13.65 A ball moving with the horizontal velocity $\mathbf{v}_0$ drops from A through the vertical distance $h_0 = 16$ in. to a frictionless floor. Knowing that the ball hits the floor at a distance $d_0 = 4$ in. from B and that the height of its first bounce is $h_1 = 9$ in., determine (*a*) the coefficient of restitution between the ball and the floor, (*b*) the length d_1 of the first bounce.

13.66 A ball moving with a horizontal velocity of magnitude $v_0 = 0.2$ m/s drops from A to a frictionless floor. Knowing that the ball hits the floor at a distance $d_0 = 60$ mm from B and that the length of its first bounce is $d_1 = 96$ mm, determine (*a*) the coefficient of restitution between the ball and the floor, (*b*) the height h_1 of the first bounce.

13.67 A sphere A of mass m hits squarely with the velocity $\mathbf{v}_0$ a sphere B of the same mass m which is hanging from an inextensible wire BC. Knowing that $\theta = 30°$ and that the coefficient of restitution between the two spheres is $e = 0.75$, determine the velocity of each sphere immediately after impact.

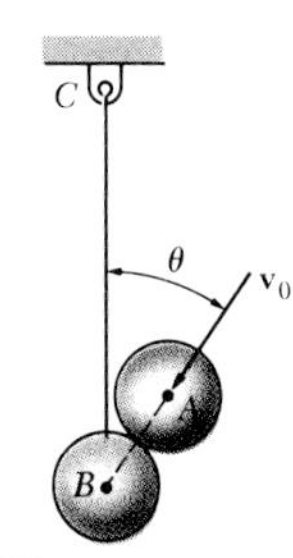

Fig. P13.67

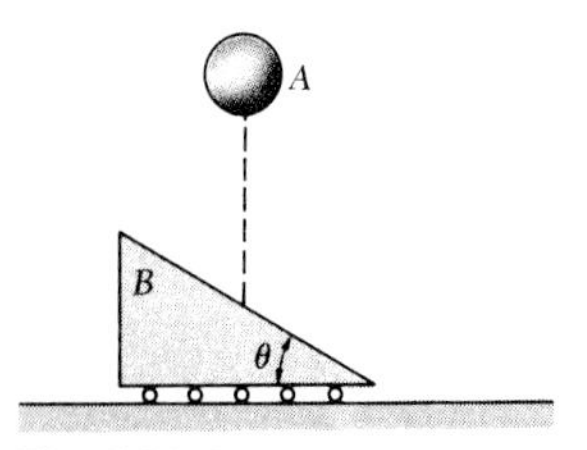

Fig. P13.68

13.68 A sphere A of mass $m_A = 2$ kg is released from rest in the position shown and strikes the frictionless, inclined surface of a wedge B of mass $m_B = 6$ kg with a velocity of magnitude $v_0 = 3$ m/s. The wedge, which is supported by rollers and may move freely in the horizontal direction, is initially at rest. Knowing that $\theta = 30°$ and that the coefficient of restitution between the sphere and the wedge is $e = 0.80$, determine the velocities of the sphere and of the wedge immediately after impact.

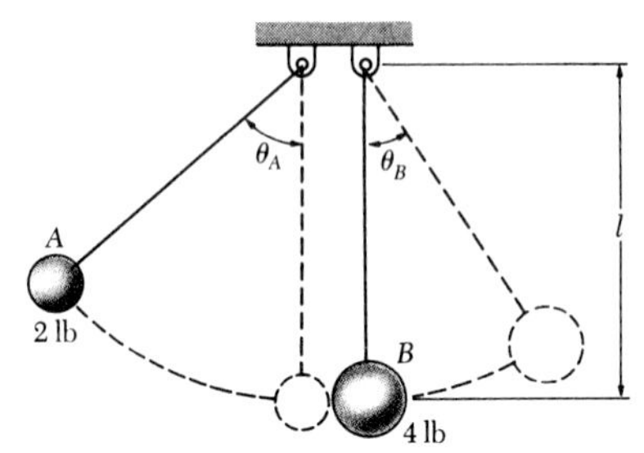

Fig. P13.69 and P13.70

13.69 A 2-lb sphere A is released from rest when $\theta_A = 50°$ and strikes a 4-lb sphere B which is at rest. Knowing that the coefficient of restitution is $e = 0.80$, determine the values of θ_A and θ_B corresponding to the highest positions to which the spheres will rise after impact.

13.70 A 2-lb sphere A is released from rest when $\theta_A = 60°$ and strikes a 4-lb sphere B which is at rest. Knowing that the velocity of sphere A is zero after impact, determine (*a*) the coefficient of restitution e, (*b*) the value of θ_B corresponding to the highest position to which sphere B will rise.

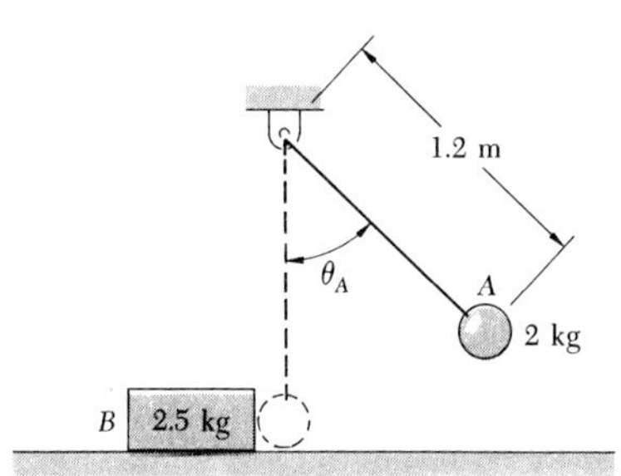

Fig. P13.71 and P13.72

13.71 A 2-kg sphere A is released from rest when $\theta_A = 60°$ and strikes a 2.5-kg block B which is at rest. It is observed that the velocity of the sphere is zero after impact and that the block moves through 1.5 m before coming to rest. Determine (*a*) the coefficient of restitution between the sphere and the block, (*b*) the coefficient of kinetic friction between the block and the floor.

13.72 A 2-kg sphere A is released from rest when $\theta_A = 90°$ and strikes a 2.5-kg block B which is at rest. Knowing that the coefficient of restitution between the sphere and the block is 0.75 and that the coefficient of kinetic friction between the block and the floor is 0.25, determine (*a*) how far block B will move, (*b*) the percentage of the initial energy lost in friction between the block and the floor.

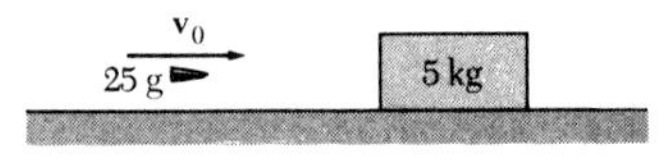

Fig. P13.73

13.73 A 25-g bullet is fired with a velocity of magnitude $v_0 = 550$ m/s into a 5-kg block of wood. Knowing that the coefficient of kinetic friction between the block and the floor is 0.30, determine (*a*) how far the block will move, (*b*) the percentage of the initial energy lost in friction between the block and the floor.

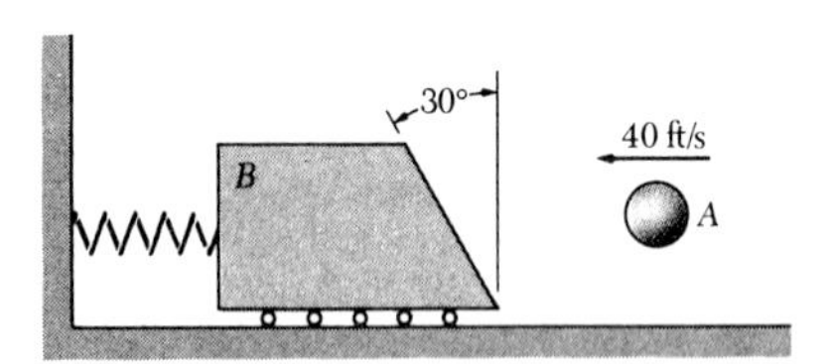

Fig. P13.74

13.74 A 2-lb sphere A is moving to the left with a velocity of 40 ft/s when it strikes the inclined surface of a 5-lb block B which is at rest. The block is supported by rollers and is attached to a spring of constant $k = 12$ lb/in. Knowing that the coefficient of restitution between the sphere and the block is $e = 0.75$ and neglecting friction, determine the maximum deflection of the spring.

CHAPTER 13 COMPUTER PROBLEMS

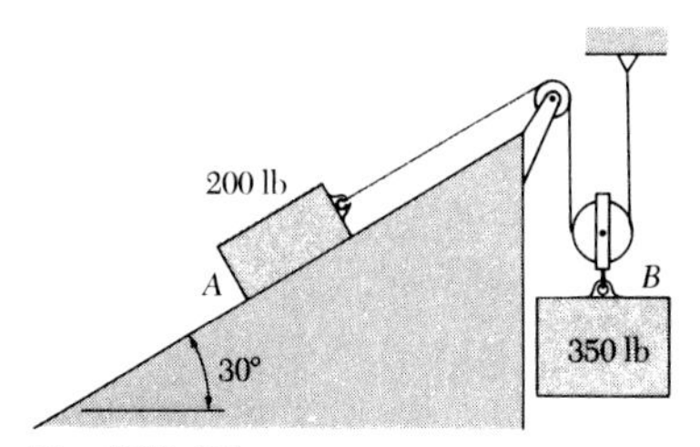

Fig. P13.C1

13.C1 The two blocks shown are released from rest. Neglect the masses of the pulleys and the effect of friction in the pulleys and between the block A and the incline.
(a) Derive an expression for the velocity of block A as a function of the distance s it travels. Plot the block's velocity as a function of s for $s = 0$ to $s = 6$ ft.
(b) From the expression in part (a) derive an expression for the position of block A as a function of time. Plot the block's position as a function of time for $s = 0$ to $s = 6$ ft.

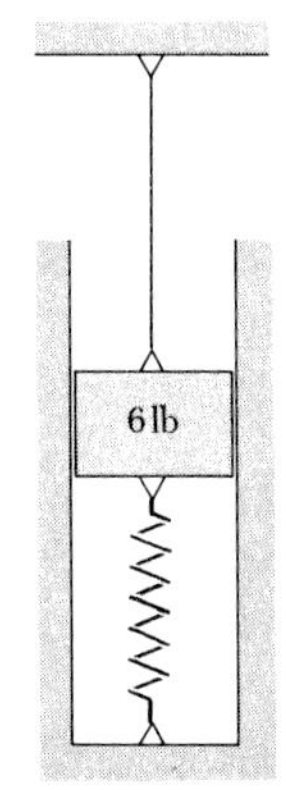

Fig. P13.C2

13.C2 A 6-lb block is attached to a cable and to a spring as shown. The constant of the spring is $k = 15$ lb/in. and the tension in the cable is 4 lb. The block is initially at rest and then the cable is cut.
(a) Derive an expression for the velocity of the block as a function of its position.
(b) Plot the block's velocity as a function of its position. What is the block's maximum velocity and maximum displacement.

13.C3 Nonlinear springs are classified as hard or soft, depending upon the curvature of their force-deflection curve (see figure). A delicate instrument having a mass of 5 kg is placed on a spring of length l so that its base is just touching the undeformed spring and then inadvertently released from that position.
Derive an expression for the velocity of the instrument as a function of the spring's deflection x, and plot the velocity as a function of x from $x = 0$ to the maximum deflection x_m of the spring for the following three cases:
(a) a linear spring with a constant of k = 3 kN/m,
(b) a hard nonlinear spring for which $F = (3\,k\text{N/m})x(1+160x^2)$,
(d) a soft nonlinear spring for which $F = (3\,k\text{N/m})x(1-160x^2)$.

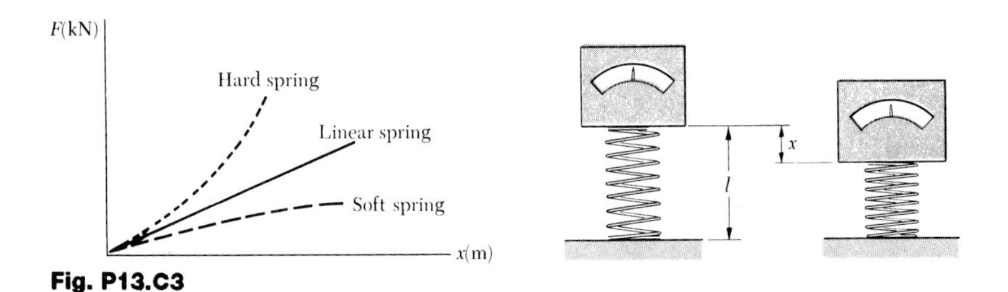

Fig. P13.C3

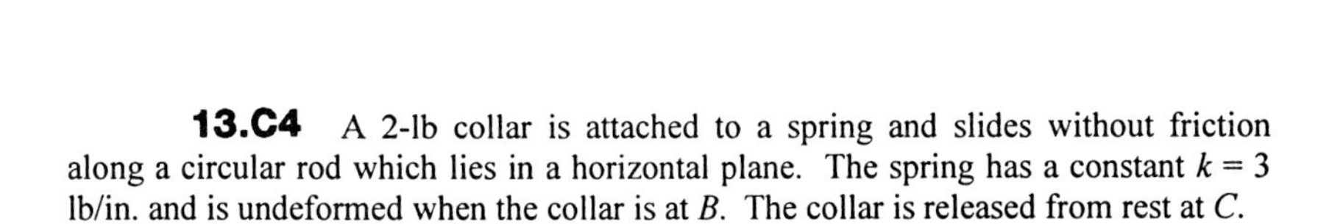

13.C4 A 2-lb collar is attached to a spring and slides without friction along a circular rod which lies in a horizontal plane. The spring has a constant $k = 3$ lb/in. and is undeformed when the collar is at B. The collar is released from rest at C.
(a) Derive an expression for the velocity of the collar as a function of the angle θ.
(b) Plot the velocity of the collar as a function of θ for $\theta = 3\pi/2$ to 2π. What is the collar's maximum velocity.

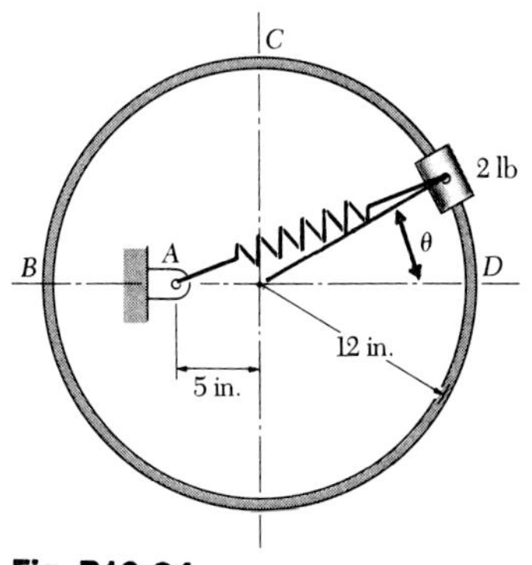

Fig. P13.C4

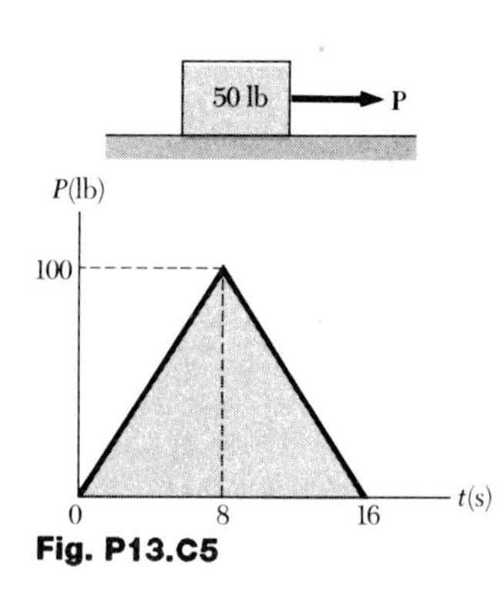

Fig. P13.C5

13.C5 A 50-lb block initially at rest is acted upon by a force **P** which varies as shown. The coefficients of friction between the block and the horizontal surface are $\mu_s = 0.5$ and $\mu_k = 0.4$. Determine the velocity of the block as a function of time. Plot the velocity of the block as a function of time for $t = 0$ to $t = 16$ s. What is the maximum velocity of the block.

CHAPTER 14
SYSTEM OF PARTICLES

SECTIONS 14.1 to 14.6

14.1 A 25-g bullet is fired in a horizontal direction through block A and becomes embedded in block B. Knowing that $v_0 = 600$ m/s and that block B starts moving with a velocity of 2 m/s, determine (a) the velocity with which block A starts moving, (b) the velocity of the bullet as it travels from block A to block B.

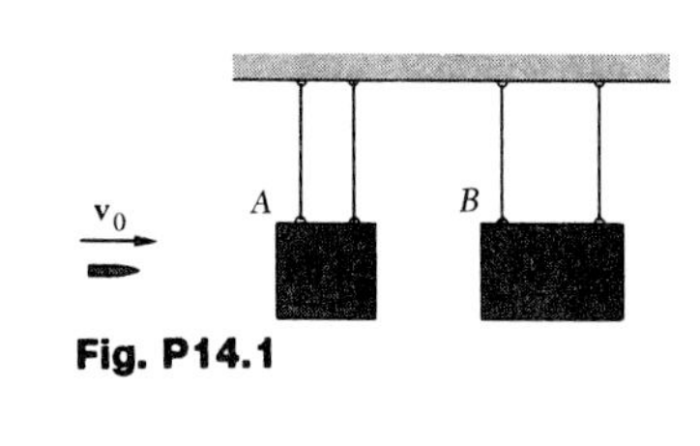

Fig. P14.1

14.2 A 40-ton boxcar A is moving with a velocity of 4.5 mi/h in a railroad switchyard when it strikes, and is automatically coupled with, a 20-ton flatcar B which carries a 30-ton trailer truck C. Both the flatcar and the truck are at rest with their brakes released. Determine the velocity of the cars and of the truck (a) immediately after the coupling of the two cars, (b) immediately after the end of the flatcar hits the truck. Neglect friction and assume the impact between the flatcar and the truck to be perfectly plastic ($e = 0$).

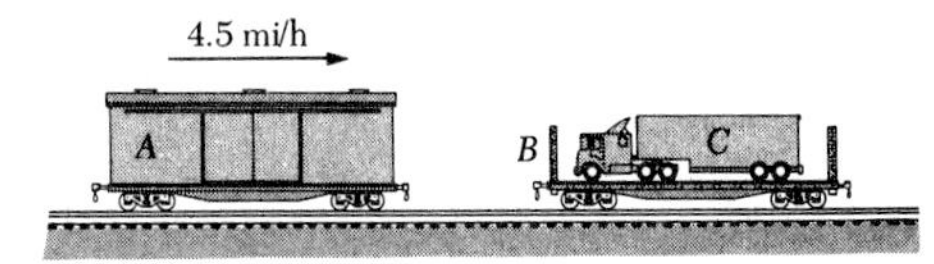

Fig. P14.2

14.3 Solve Prob. 14.2, assuming the impact between the flatcar and the truck to be perfectly elastic ($e = 1$).

14.4 Three identical freight cars have the velocities indicated. Assuming that car B is first hit by car C, determine the velocity of each car after all collisions have taken place, if (a) all three cars get automatically coupled, (b) cars A and B get automatically and tightly coupled, while cars B and C bounce off each other with a coefficient of restitution $e = 1$.

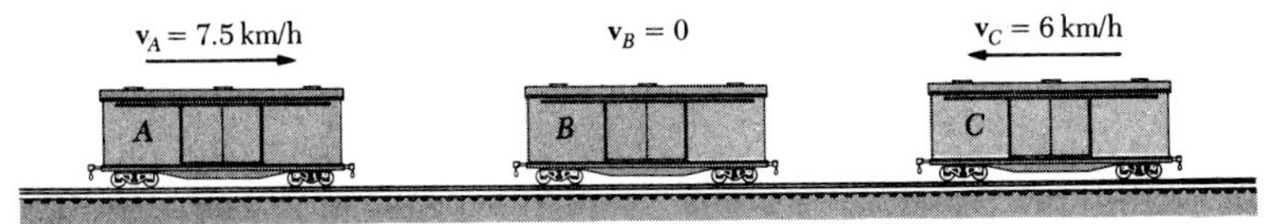

Fig. P14.4

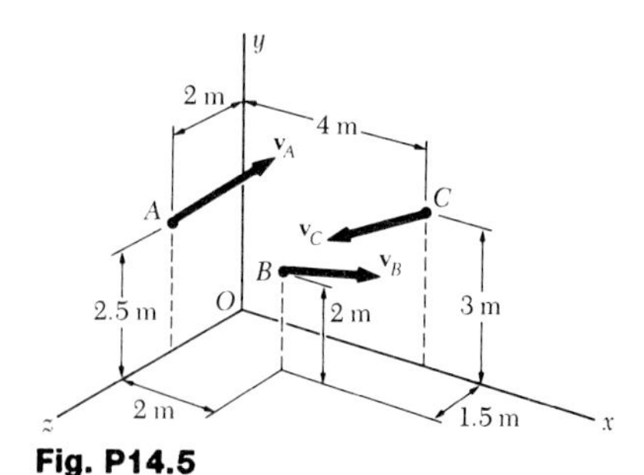

Fig. P14.5

14.5 A system consists of three particles A, B, and C. We know that $m_A = 3$ kg, $m_B = 4$ kg, and $m_C = 5$ kg and that the velocities of the particles expressed in m/s are, respectively, $\mathbf{v}_A = 2\mathbf{i} + 3\mathbf{j} - 2\mathbf{k}$, $\mathbf{v}_B = v_x\mathbf{i} + 2\mathbf{j} + v_z\mathbf{k}$, and $\mathbf{v}_C = -3\mathbf{i} - 2\mathbf{j} + \mathbf{k}$. Determine (*a*) the components v_x and v_z of the velocity of particle B for which the angular momentum $\mathbf{H}_O$ of the system about O is parallel to the x axis, (*b*) the corresponding value of $\mathbf{H}_O$.

14.6 For the system of particles of Prob. 14.5, determine (*a*) the components v_x and v_z of the velocity of particle B for which the angular momentum $\mathbf{H}_O$ of the system about O is parallel to the z axis, (*b*) the corresponding value of $\mathbf{H}_O$.

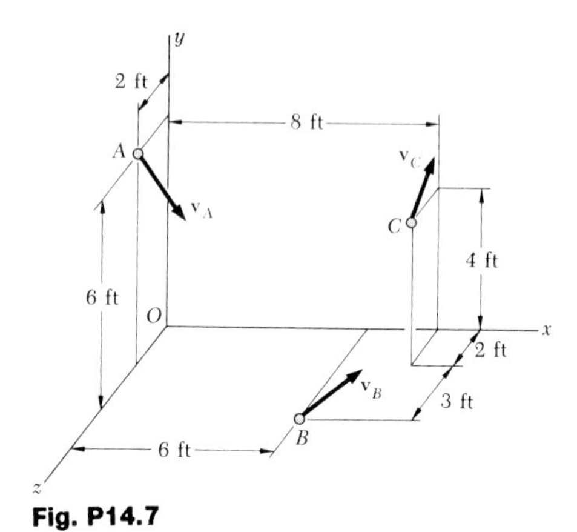

Fig. P14.7

14.7 A system consists of three particles A, B, and C. We know that $W_A = 2$ lb, $W_B = 4$ lb, and $W_C = 6$ lb and that the velocities of the particles expressed in ft/s are, respectively, $\mathbf{v}_A = 6\mathbf{i} - 4\mathbf{j} + 8\mathbf{k}$, $\mathbf{v}_B = 8\mathbf{i} + 6\mathbf{j}$, and $\mathbf{v}_C = 4\mathbf{i} + 10\mathbf{j} - 6\mathbf{k}$. Determine the angular momentum $\mathbf{H}_O$ of the system about O.

14.8 For the system of particles of Prob. 14.7, determine (*a*) the position vector $\bar{\mathbf{r}}$ of the mass center G of the system, (*b*) the linear momentum $m\bar{\mathbf{v}}$ of the system, (*c*) the angular momentum $\mathbf{H}_G$ of the system about G.

14.9 A 600-lb space vehicle traveling with a velocity $\mathbf{v}_0 = (1200 \text{ ft/s})\mathbf{k}$ passes through the origin O at $t = 0$. Explosive charges then separate the vehicle into three parts A, B, and C, weighing, respectively, 100 lb, 200 lb, and 300 lb. Knowing that at $t = 2$ s, the positions of parts B and C are observed to be B (600, 1320, 3240) and C (−480, −960, 1920), where the coordinates are expressed in feet, determine the corresponding position of part A. Neglect the effect of gravity.

14.10 An airliner weighing 50 tons and flying due east at an altitude of 28,000 ft and a speed of 540 mi/h suddenly explodes, breaking into three fragments A, B, and C. Fragment A, weighing 25 tons, and fragment B, weighing 15 tons, are found in a wooded area at the points shown in the figure. Knowing that the explosion occurred when the airliner was directly above point O and assuming that all three fragments hit the ground at the same time, determine the location of fragment C. Neglect the resistance of the air.

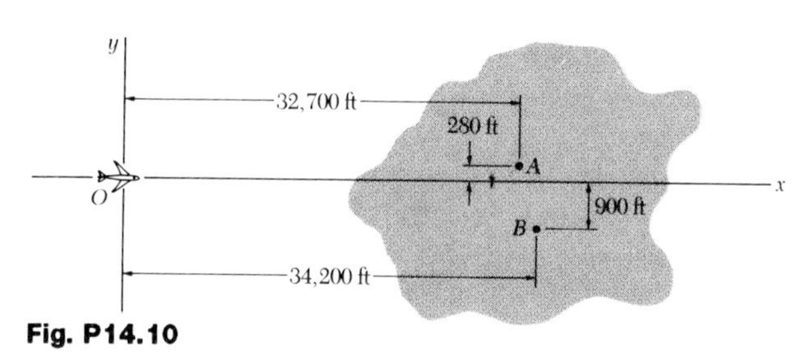

Fig. P14.10

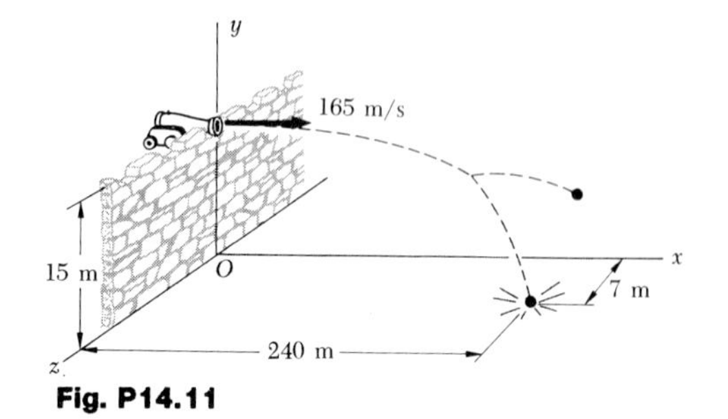

Fig. P14.11

14.11 Two 15-kg cannon balls are chained together and fired horizontally with a velocity of 165 m/s from the top of a 15-m wall. The chain breaks during the flight of the cannon balls and one of them strikes the ground at $t = 1.5$ s, at a distance of 240 m from the foot of the wall, and 7 m to the right of the line of fire. Determine the position of the other cannon ball at that instant. Neglect the resistance of the air.

14.12 Solve Prob. 14.11, if the cannon ball which first strikes the ground has a mass of 12 kg, and the other a mass of 18 kg. Assume that the time of flight and the point of impact of the first cannon ball remain the same.

14.13 A 500-lb space vehicle is traveling with a velocity $\mathbf{v}_0 = (1500 \text{ ft/s})\mathbf{i}$ when explosive charges separate it into three parts A, B, and C, weighing, respectively, 120 lb, 180 lb, and 200 lb. Knowing that immediately after the explosion, the velocity of part A is $\mathbf{v}_A = (2000 \text{ ft/s})\mathbf{i} - (200 \text{ ft/s})\mathbf{j} + (300 \text{ ft/s})\mathbf{k}$ and that the velocity of part B is $\mathbf{v}_B = (1000 \text{ ft/s})\mathbf{i} + (300 \text{ ft/s})\mathbf{j} - (500 \text{ ft/s})\mathbf{k}$, determine the corresponding velocity of part C.

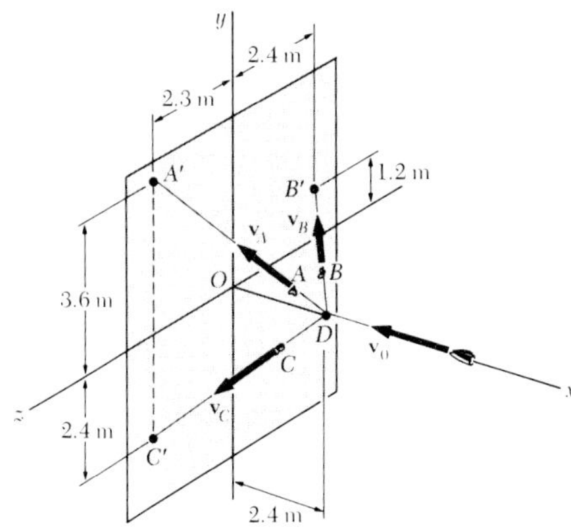

Fig. P14.14

14.14 A 1-kg shell is moving in a direction perpendicular to a wall when at point D, it explodes into three fragments A, B, and C, respectively, of mass 300 g, 300 g, and 400 g. Knowing that the fragments hit the wall at the points indicated and that $v_A = 490$ m/s, determine the velocity $\mathbf{v}_0$ of the shell before it exploded.

14.15 An archer hits a game bird flying in a horizontal straight line 9 m above the ground with a 32-g wooden arrow. Knowing that the arrow strikes the bird from behind with a velocity of 110 m/s at an angle of 30° with the vertical, and that the bird falls to the ground in 1.5 s and 15 m beyond the point where it was hit, determine (*a*) the mass of the bird, (*b*) the speed at which it was flying when it was hit.

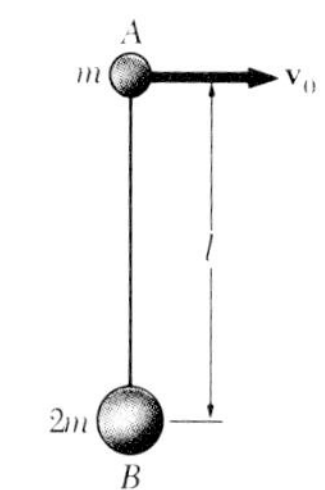

Fig. P14.16

14.16 Two small spheres A and B, respectively of mass m and $2m$, are connected by a rigid rod of length l and negligible mass. The two spheres are resting on a horizontal, frictionless surface when A is suddenly given the velocity $\mathbf{v}_0 = v_0\mathbf{i}$. Determine (*a*) the linear momentum of the system and its angular momentum about its mass center G, (*b*) the velocities of A and B after the rod AB has rotated through 90°, (*c*) the velocities of A and B after the rod AB has rotated through 180°.

SECTIONS 14.7 to 14.9

14.17 In Prob. 14.2, determine the energy lost (*a*) in the coupling of the two cars, (*b*) as the end of the flatcar hits the truck.

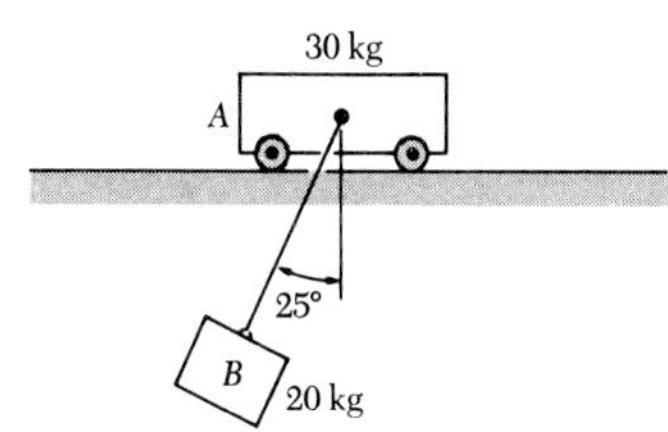

Fig. P14.19

14.18 A space vehicle which is drifting in space consists of two parts A and B connected by explosive bolts. Parts A and B compress four springs, each of which has a potential energy of 120 ft·lb. When the bolts are exploded the springs expand, causing parts A and B to move away from each other. Determine the relative velocity of B with respect to A, knowing that (*a*) each part weighs 2000 lb, (*b*) part A weighs 3000 lb and part B 1000 lb.

14.19 A 20-kg block B is suspended from a 2-m cord attached to a 30-kg cart A, which may roll freely on a frictionless, horizontal track. If the system is released from rest in the position shown, determine the velocities of A and B as B passes directly under A.

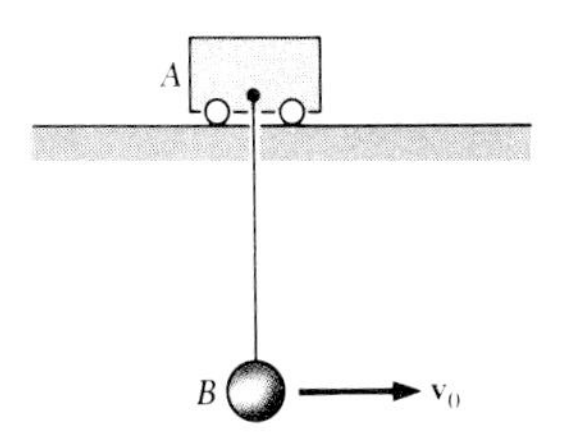

Fig. P14.20 and P14.21

14.20 Ball B, of mass m_B, is suspended from a cord of length l attached to a cart A, of mass m_A, which may roll freely on a frictionless, horizontal track. If the ball is given an initial horizontal velocity $\mathbf{v}_0$ while the cart is at rest, determine (*a*) the maximum vertical distance h through which B will rise, (*b*) the velocity of B as it reaches its maximum elevation. (*Hint.* At that instant, the velocity of B relative to A is zero.) Discuss the particular case when $m_A \gg m_B$ and when $m_A \ll m_B$.

14.21 Ball B is suspended from a cord of length l attached to a cart A, which may roll freely on a frictionless, horizontal track. The ball and the cart have the same mass m. If the ball is given an initial horizontal velocity $\mathbf{v}_0$ while the cart is at rest, describe the subsequent motion of the system, specifying the velocities of A and B for the following successive values of the angle θ (assumed positive counterclockwise) that the cord will form with the vertical: (*a*) $\theta = \theta_{max}$, (*b*) $\theta = 0$, (*c*) $\theta = \theta_{min}$.

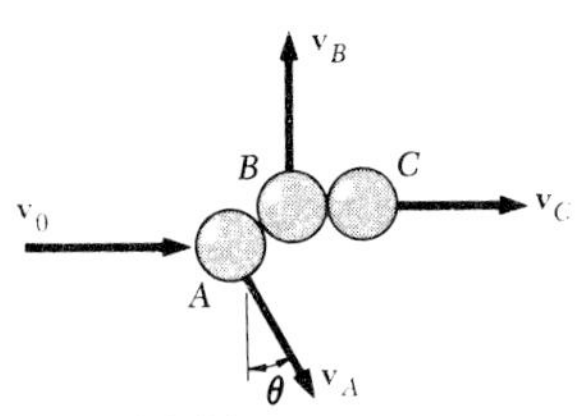

Fig. P14.22

14.22 In a game of billiards, ball A is moving with the velocity $\mathbf{v}_0 = (4 \text{ m/s})\mathbf{i}$ when it strikes balls B and C which are at rest side by side. After the collision, the three balls are observed to move in the directions shown, with $\theta = 25°$. Assuming frictionless surfaces and perfectly elastic impacts (i.e., conservation of energy), determine the magnitudes of the velocities v_A, v_B, and v_C.

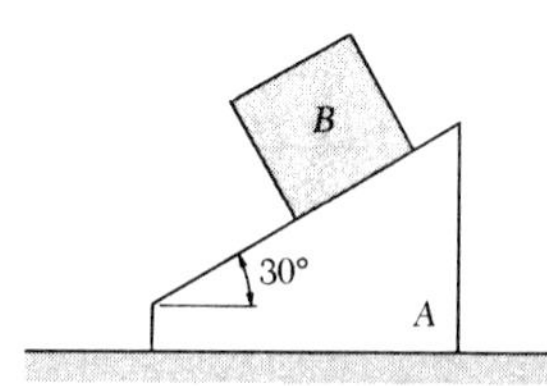

Fig. P14.23

14.23 A 12-lb block B starts from rest and slides on the 30-lb wedge A, which is supported by a horizontal surface. Neglecting friction, determine (a) the velocity of B relative to A after it has slid 3 ft down the inclined surface of the wedge, (b) the corresponding velocity of A.

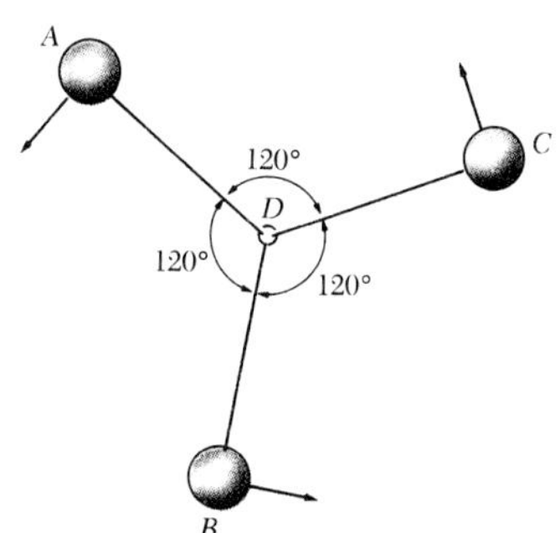

Fig. P14.24

14.24 Three small spheres A, B, and C, each of mass m, are connected to a small ring D by means of three inextensible, inelastic cords of length l which are equally spaced. The spheres may slide freely on a frictionless horizontal surface and are rotating initially with a speed v_0 about ring D which is at rest. Suddenly cord CD breaks. After the other two cords have again become taut, determine (a) the speed of ring D, (b) the relative speed with which spheres A and B rotate about D, (c) the fraction of the energy of the original system which is dissipated when cords AD and BD again become taut.

14.25 A 600-lb space vehicle traveling with a velocity $\mathbf{v}_0 = (1200 \text{ ft/s})\mathbf{k}$ passes through the origin O. Explosive charges then separate the vehicle into three parts A, B, and C, weighing, respectively, 100 lb, 200 lb, and 300 lb. Knowing that shortly thereafter the positions of the three parts are, respectively, $A(240, 240, 2160)$, $B(600, 1320, 3240)$, and $C(-480, -960, 1920)$, where the coordinates are expressed in feet, that the velocity of B is $\mathbf{v}_B = (400 \text{ ft/s})\mathbf{i} + (880 \text{ ft/s})\mathbf{j} + (1760 \text{ ft/s})\mathbf{k}$, and that the x component of the velocity of C is -320 ft/s, determine the velocity of part A.

14.26 Three small identical spheres A, B, and C are attached to three strings, 150 mm long, which are tied to a ring G. The ring has a velocity $\mathbf{v}_0 = (0.3 \text{ m/s})\mathbf{i}$ and moves along the x axis. The three spheres are rotating counterclockwise about G with a relative velocity of 0.6 m/s. Suddenly ring G breaks and the three spheres move freely in the horizontal xy plane, with A and B following paths parallel to the y axis and C a path parallel to the x axis. Determine (a) the velocity of each sphere, (b) the distance d.

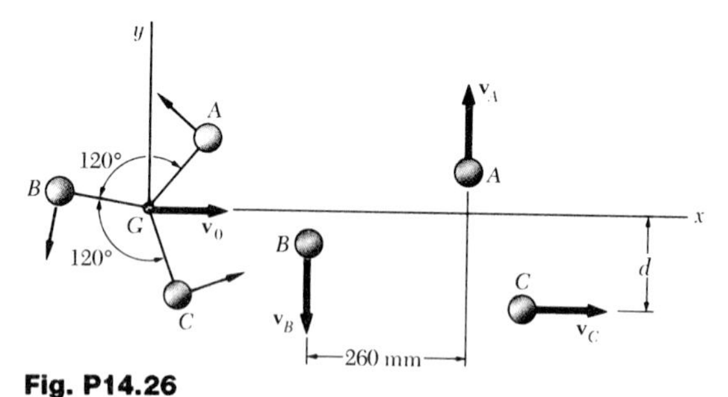

Fig. P14.26

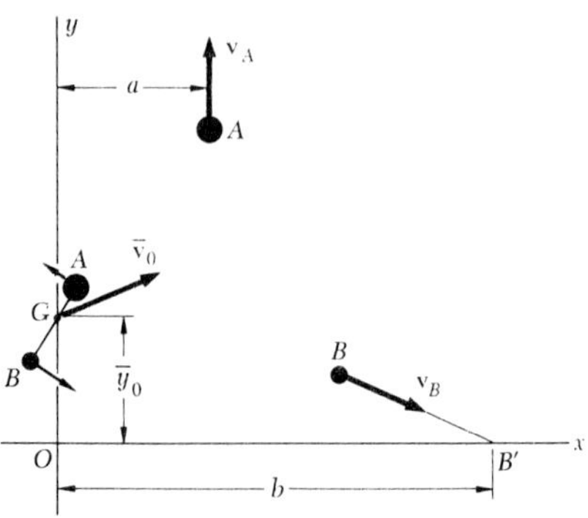

Fig. P14.27

14.27 Two small disks A and B, of mass 3 kg and 1.5 kg, respectively, may slide on a horizontal, frictionless surface. They are connected by a cord, 600 mm long, and spin counterclockwise about their mass center G at the rate of 10 rad/s. At $t = 0$, the coordinates of G are $\bar{x}_0 = 0$, $\bar{y}_0 = 2$ m, and its velocity is $\bar{\mathbf{v}}_0 = (1.2 \text{ m/s})\mathbf{i} + (0.96 \text{ m/s})\mathbf{j}$. Shortly thereafter the cord breaks; disk A is then observed to move along a path parallel to the y axis and disk B along a path which intersects the x axis at a distance $b = 7.5$ m from O. Determine (a) the velocities of A and B after the cord breaks, (b) the distance a from the y axis to the path of A.

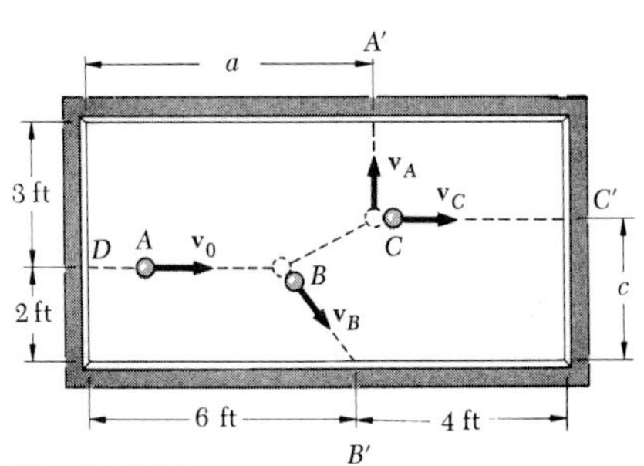

Fig. P14.28

14.28 In a game of billiards, ball A is given an initial velocity $\mathbf{v}_0$ along line DA parallel to the axis of the table. It hits ball B and then ball C, which are both at rest. Balls A and C are observed to hit the sides of the table squarely at points A' and C', respectively, and ball B to hit the side obliquely at B'. Knowing that $v_0 = 12$ ft/s, $v_C = 7$ ft/s, and $c = 3$ ft., determine (*a*) the velocities $\mathbf{v}_A$ and $\mathbf{v}_B$ of balls A and B, (*b*) the point A' where ball A hits the side of the table. Assume frictionless surfaces and perfectly elastic impacts (i.e., conservation of energy).

14.29 For the game of billiards of Prob. 14.28, it is now assumed that $v_0 = 10$ ft/s, $v_A = 4$ ft/s, and $a = 7$ ft. Determine (*a*) the velocities $\mathbf{v}_B$ and $\mathbf{v}_C$ of balls B and C, (*b*) the point C' where ball C hits the side of the table.

14.30 In a game of billiards, ball A is moving with the velocity $\mathbf{v}_0 = (3 \text{ m/s})\mathbf{i}$ when it strikes balls B and C which are at rest side by side. After the collision, A is observed to move with the velocity $\mathbf{v}_A = (1.176 \text{ m/s})\mathbf{i} - (1.368 \text{ m/s})\mathbf{j}$, while B and C move in the directions shown. Determine (*a*) the magnitudes of the velocities $\mathbf{v}_B$ and $\mathbf{v}_C$, (*b*) the percentage of the initial energy lost in the collision.

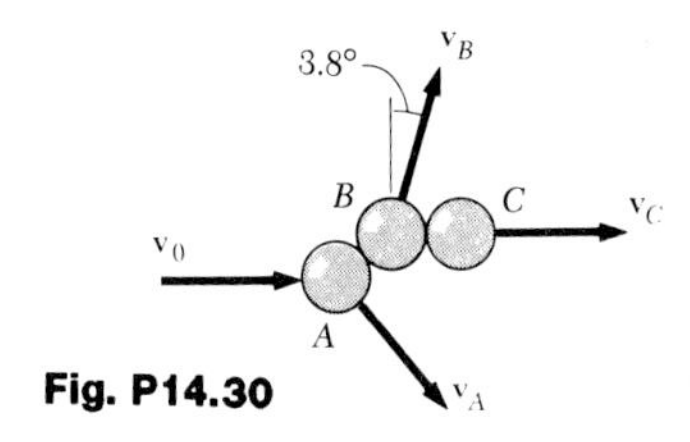

Fig. P14.30

14.31 Three identical freight cars are located on the same track. Car A is moving to the right with a velocity $\mathbf{v}_0$, while cars B and C are at rest. Determine the velocity of each car after all collisions have taken place, assuming that cars A and B bounce off each other with a coefficient of restitution $e = 1$ and that cars B and C (*a*) also bounce off each other with $e = 1$, (*b*) get automatically coupled as they hit each other, (*c*) were already tightly coupled when A hit B.

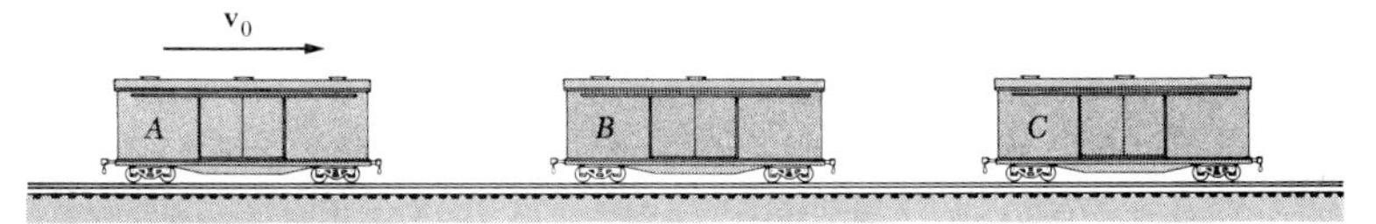

Fig. P14.31

14.32 When the cord connecting particles A and B is severed, the compressed spring causes the particles to fly apart (the spring is not connected to the particles). The potential energy of the compressed spring is known to be 60 J and the assembly has an initial velocity $\mathbf{v}_0$ as shown. If the cord is severed when $\theta = 25°$, determine the resulting velocity of each particle.

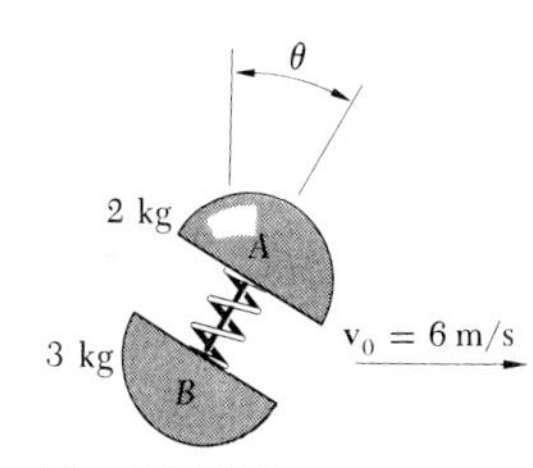

Fig. P14.32

SECTIONS 14.10 to 14.12

14.33 A stream of water of cross-sectional area A and velocity $\mathbf{v}_1$ strikes a plate which is held motionless by a force $\mathbf{P}$. Determine the magnitude of $\mathbf{P}$, knowing that $A = 600 \text{ mm}^2$, $v_1 = 30$ m/s, and $V = 0$.

14.34 A stream of water of cross-sectional area A and velocity $\mathbf{v}_1$ strikes a plate which moves to the right with a velocity $\mathbf{V}$. Determine the magnitude of $\mathbf{V}$, knowing that $A = 400 \text{ mm}^2$, $v_1 = 35$ m/s, and $P = 360$ N.

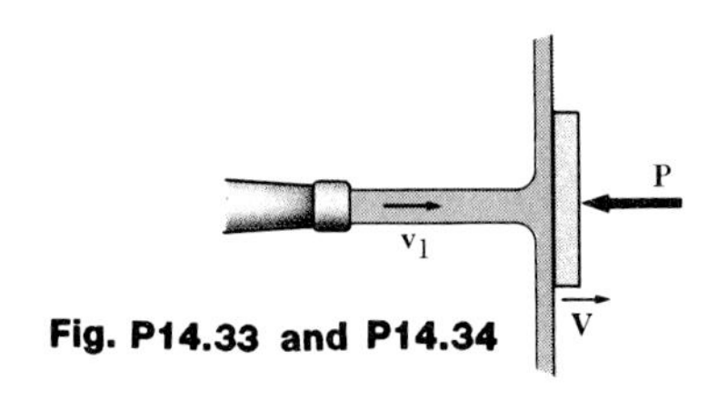

Fig. P14.33 and P14.34

14.35 Water flows in a continuous sheet from between two plates A and B with a velocity $\mathbf{v}$ of magnitude 80 ft/s. The stream is split into two parts by a smooth horizontal plate C. Determine the rates of flow Q_1 and Q_2 in each of the two resulting streams, knowing that $\theta = 30°$ and that the total force exerted by the streams on the plate is a 75-lb vertical force ($1 \text{ ft}^3 = 7.48$ gal).

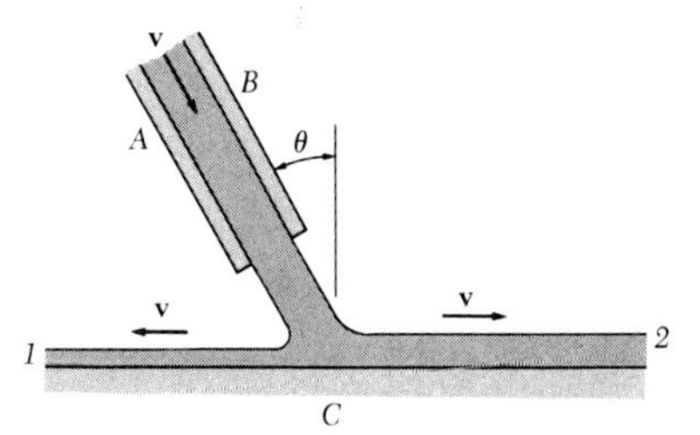

Fig. P14.35

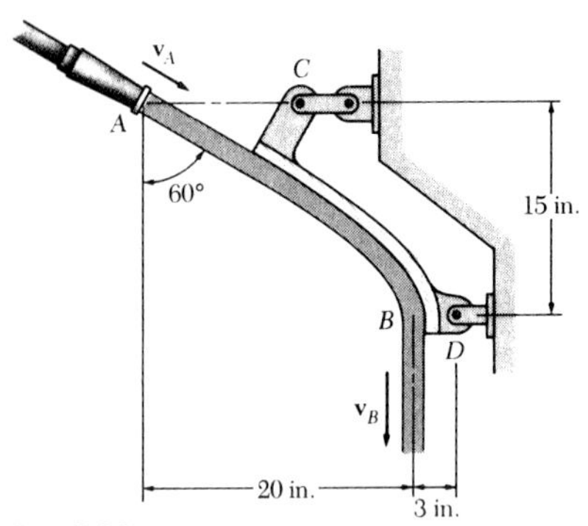

Fig. P14.36

14.36 The nozzle shown discharges water at the rate of 300 gal/min. Knowing that at both A and B the stream of water moves with a velocity of magnitude 80 ft/s and neglecting the weight of the vane, determine the components of the reactions at C and D ($1\ \text{ft}^3 = 7.48$ gal).

14.37 While cruising in level flight at a speed of 600 mi/h, a jet airplane scoops in air at the rate of 170 lb/s and discharges it with a velocity of 2000 ft/s relative to the airplane. Determine the total drag due to air friction on the airplane.

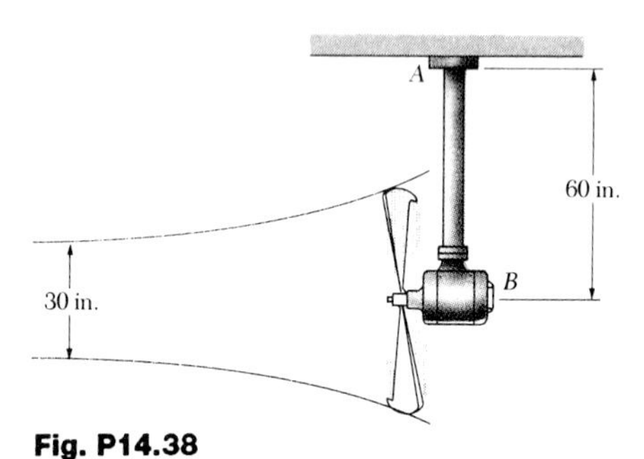

Fig. P14.38

14.38 For the ceiling-mounted fan shown, determine the maximum allowable air velocity in the slipstream if the bending moment in the supporting rod AB is not to exceed 80 lb·ft. Assume $\gamma = 0.076\ \text{lb/ft}^3$ for air and neglect the approach velocity of the air.

14.39 The helicopter shown has a mass of 10 Mg. Assuming $\rho = 1.21\ \text{kg/m}^3$ for air, determine the downward air speed the helicopter produces in its 16-m-diameter slipstream while hovering in midair with a combined payload and fuel load of 4000 kg.

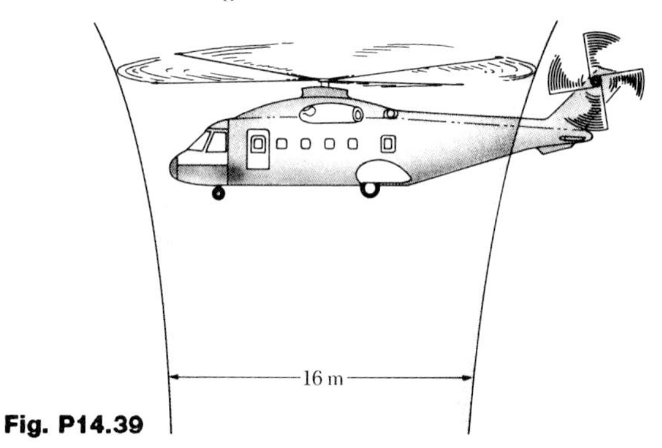

Fig. P14.39

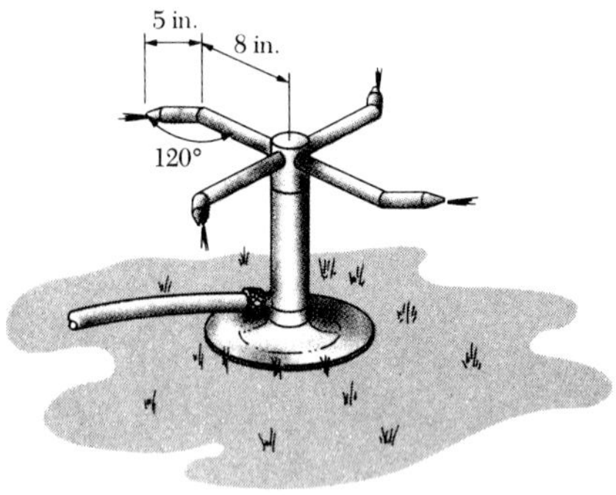

14.40 A garden sprinkler has four rotating arms, each of which consists of two horizontal straight sections of pipe forming an angle of 120°. Each arm discharges water at the rate of 4 gal/min with a velocity of 50 ft/s relative to the arm. Knowing that the friction between the moving and stationary parts of the sprinkler is equivalent to a couple of magnitude $M = 0.250$ lb·ft, determine the constant rate at which the sprinkler rotates ($1\ \text{ft}^3 = 7.48$ gal).

5 in.
8 in.
120°

Fig. P14.40

14.41 The ends of a chain lie in piles at A and C. When released from rest at time $t = 0$, the chain moves over the pulley at B, which has a negligible mass. Denoting by L the length of chain connecting the two piles and neglecting friction, determine the speed v of the chain at time t.

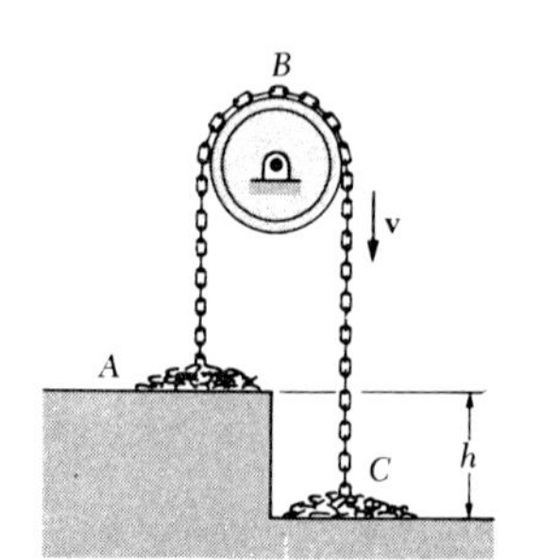

Fig. P14.41 and P14.42

14.42 For the chain of Prob. 14.41, it is known that $h = 300$ mm, and that the length of chain connecting the two piles is $L = 4$ m. Neglecting friction, determine at $t = 2$ s (a) the speed v of the chain, (b) the length of chain which has been transferred from pile A to pile C.

14.43 The acceleration of a rocket is observed to be 30 m/s^2 at $t = 0$, as it is fired vertically from the ground, and 350 m/s^2 at $t = 80$ s. Knowing that the fuel is consumed at the rate of 10 kg/s, determine (*a*) the initial mass of the rocket, (*b*) the relative velocity with which the fuel is ejected.

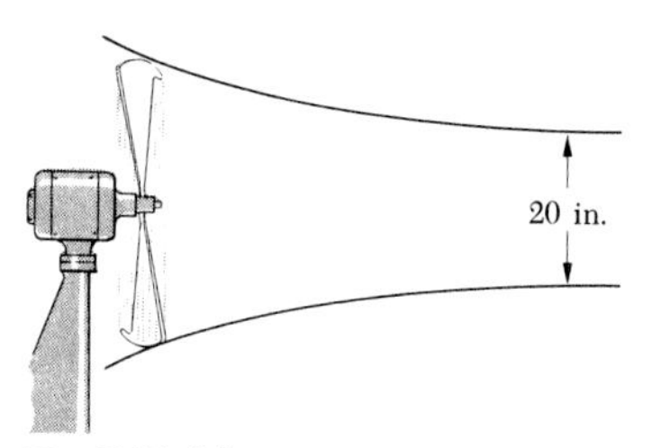

Fig. P14.44

14.44 The slipstream of a fan has a diameter of 20 in. and a velocity of 35 ft/s relative to the fan. Assuming $\gamma = 0.076$ lb/ft^3 for air and neglecting the velocity of approach of the air, determine the force required to hold the fan motionless.

14.45 The stream of water shown flows at the rate of 250 gal/min and moves with a velocity of magnitude 120 ft/s at both A and B. The vane is supported by a pin connection at C and by a load cell at D which can exert only a vertical force. Neglecting the weight of the vane, determine the reactions at C and D (1 ft^3 = 7.48 gal).

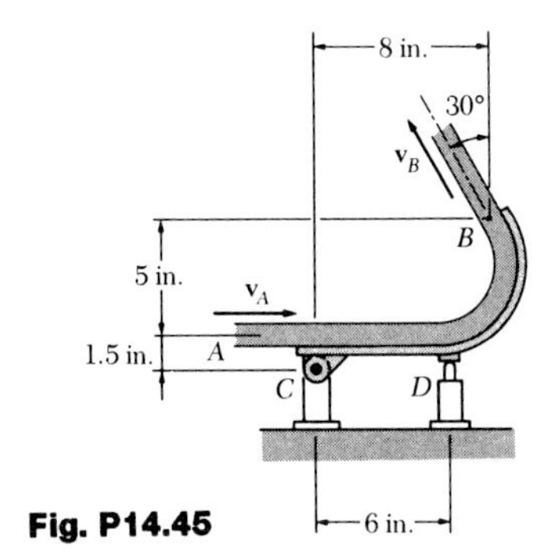

Fig. P14.45

14.46 A spacecraft is launched vertically by a two-stage rocket. When the speed is 16×10^3 km/h, the first-stage casing is released and the second-stage rocket is fired. Fuel is consumed at the rate of 100 kg/s and ejected with a relative velocity of 2500 m/s. Knowing that the combined mass of the second-stage rocket and spacecraft is 10 Mg, including 8.5 Mg of fuel, determine (*a*) the maximum speed attained by the spacecraft, (*b*) the distance between the spacecraft and the first-stage casing as the last particle of fuel is being expelled by the second-stage rocket.

14.47 Gravel falls with practically zero velocity onto a conveyor belt at the constant rate $q = dm/dt$. (*a*) Determine the magnitude of the force $\mathbf{P}$ required to maintain a constant belt speed v. (*b*) Show that the kinetic energy acquired by the gravel in a given time interval is equal to half the work done in that interval by the force $\mathbf{P}$. Explain what happens to the other half of the work done by $\mathbf{P}$.

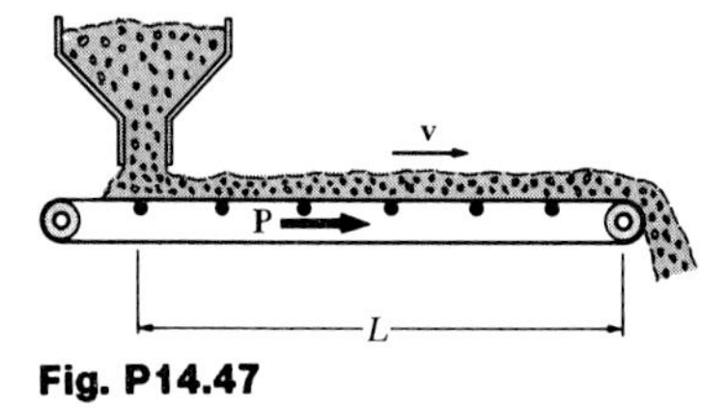

Fig. P14.47

14.48 A space vehicle describing a circular orbit at a speed of 24×10^3 km/h releases at its front end a capsule which has a gross mass of 500 kg, including 375 kg of fuel. If the fuel is consumed at the constant rate of 15 kg/s and ejected with a relative velocity of 2500 m/s, determine (*a*) the tangential acceleration of the capsule as the engine is fired, (*b*) the maximum speed attained by the capsule.

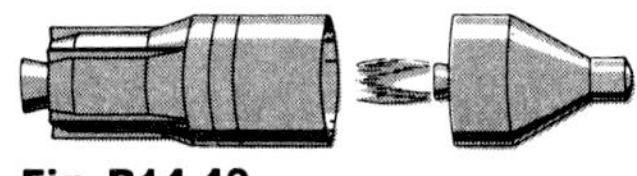
Fig. P14.48

CHAPTER 15
KINEMATICS OF RIGID BODIES

SECTIONS 15.1 to 15.4

15.1 The motion of an oscillating crank is defined by the relation $\theta = \theta_0 \sin(2\pi t/T)$, where θ is expressed in radians and t in seconds. Knowing that $\theta_0 = 1.2$ rad and $T = 0.5$ s, determine the maximum angular velocity and the maximum angular acceleration of the crank.

15.2 The motion of a disk rotating in an oil bath is defined by the relation $\theta = \theta_0(1 - e^{-t/4})$, where θ is expressed in radians and t in seconds. Knowing that $\theta_0 = 0.80$ rad, determine the angular coordinate, velocity, and acceleration of the disk when (*a*) $t = 0$, (*b*) $t = 4$ s, (*c*) $t = \infty$.

15.3 The bent rod *ABCD* rotates about a line joining points *A* and *D* with a constant angular velocity of 95 rad/s. Knowing that at the instant considered the velocity of corner *C* is downward, determine the velocity and acceleration of corner *B*.

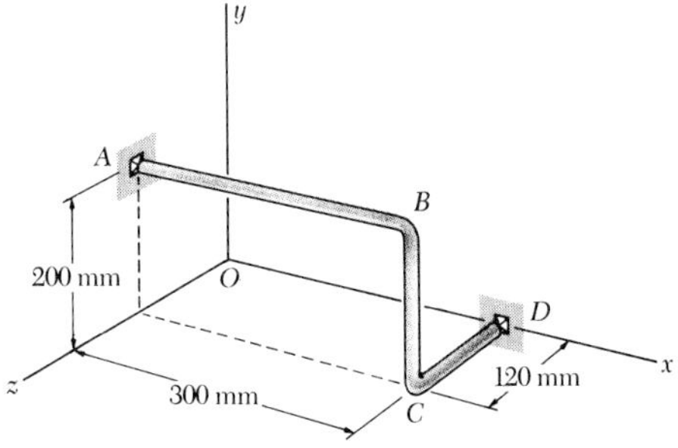

Fig. P15.3

15.4 A triangular plate and two rectangular plates are welded to each other and to the straight rod *AB*. The entire welded unit rotates about the axis *AB* with a constant angular velocity of 5 rad/s. Knowing that at the instant considered the velocity of corner *E* is downward, determine the velocity and acceleration of corner *D*.

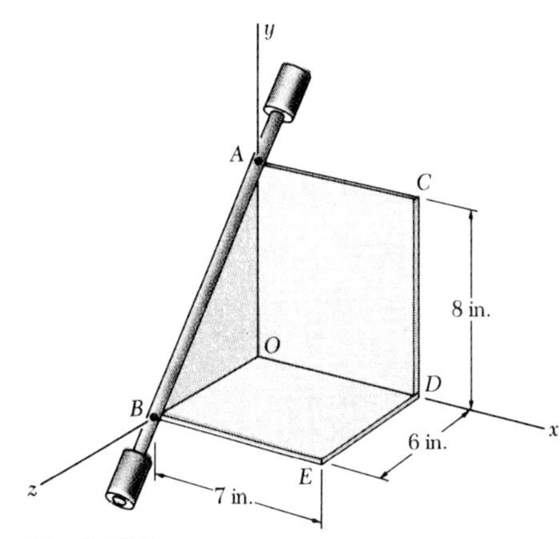

Fig. P15.4

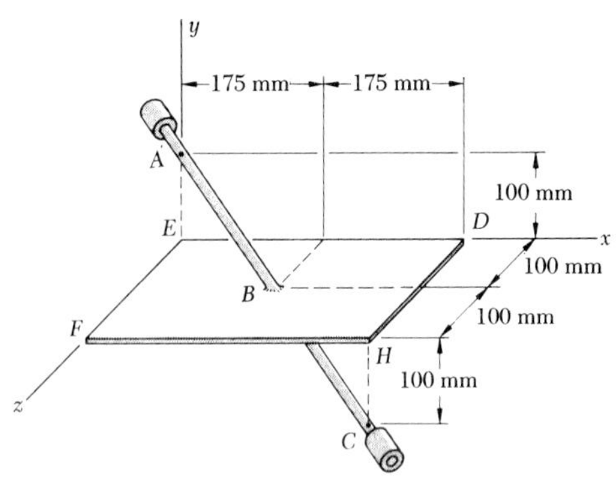

Fig. P15.5

15.5 The assembly shown consists of the straight rod *ABC* which passes through and is welded to the rectangular plate *DEFH*. The assembly rotates about the axis *AC* with a constant angular velocity of 9 rad/s. Knowing that the motion when viewed from *C* is counterclockwise, determine the velocity and acceleration of corner *F*.

15.6 It is known that the static-friction force between the small block *B* and the plate will be exceeded and that the block will start sliding on the plate when the total acceleration of the block reaches 3 m/s^2. If the plate starts from rest at $t = 0$ and is accelerated at the constant rate of 4 rad/s^2, determine the time t and the angular velocity of the plate when the block starts sliding, assuming $r = 200$ mm.

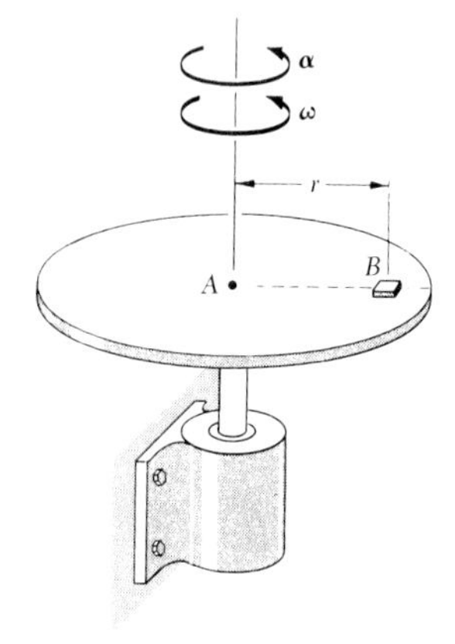

Fig. P15.6 and P15.7

15.7 A small block *B* rests on a horizontal plate which rotates about a fixed axis. The plate starts from rest at $t = 0$ and is accelerated at the constant rate of 0.5 rad/s^2. Knowing that $r = 200$ mm, determine the magnitude of the total acceleration of the block when (*a*) $t = 0$, (*b*) $t = 1$ s, (*c*) $t = 2$ s.

15.8 The belt sander shown is initially at rest. If the driving drum *A* has a constant angular acceleration of 120 rad/s^2 counterclockwise, determine the acceleration of the belt at point *C* and at the work station *D*, two seconds after the sander has been turned on.

15.9 The rated speed of drum *A* of the belt sander shown is 3450 rpm. When the power is turned off it is observed that the sander coasts from its rated speed to rest in 5 s. Assuming uniformly decelerated motion, determine the velocity and acceleration of point *C* of the belt (*a*) immediately before the power is turned off, (*b*) 4.5 s later.

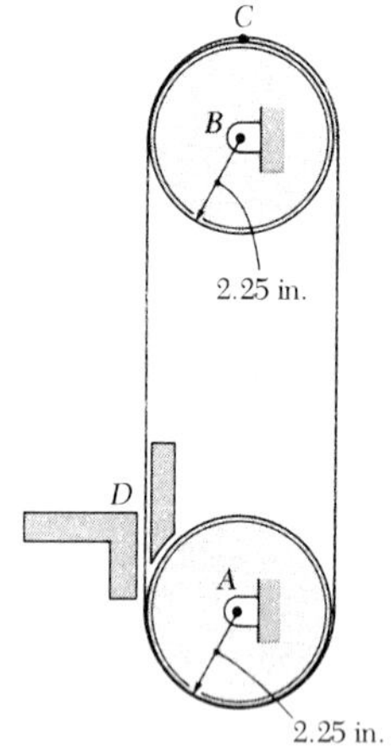

Fig. P15.8 and P15.9

15.10 Ring *C* has an inside radius of 72 mm and an outside radius of 76 mm and is positioned between two wheels *A* and *B*, each of 30-mm outside radius. Knowing that wheel *A* rotates with a constant angular velocity of 400 rpm and that no slipping occurs, determine (*a*) the angular velocity of ring *C* and of wheel *B*, (*b*) the acceleration of the points of *A* and *B* which are in contact with *C*.

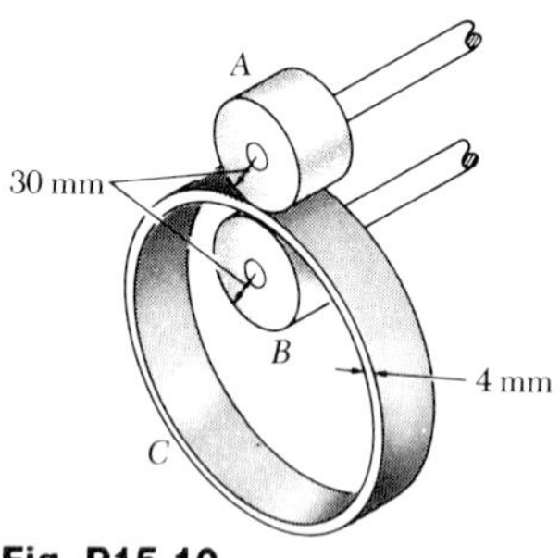

Fig. P15.10

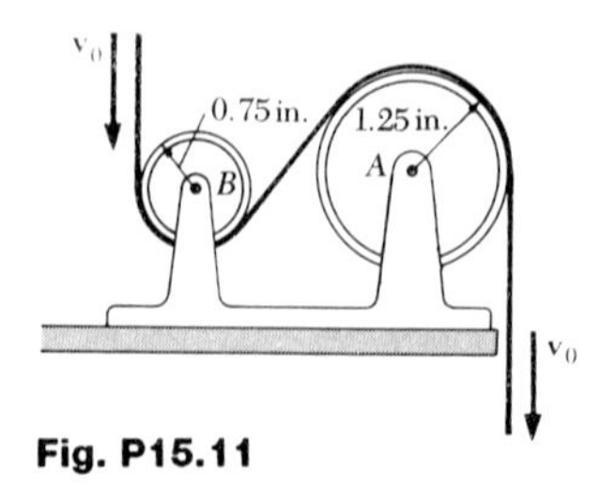

Fig. P15.11

15.11 A computer tape moves over two drums. During a 3-s interval the speed of the tape is increased uniformly from $v_0 = 2$ ft/s to $v_1 = 5$ ft/s. Knowing that the tape does not slip on the drums, determine (*a*) the angular acceleration of drum *A*, (*b*) the number of revolutions executed by drum *A* during the 3-s interval.

15.12 Solve Prob. 15.11, considering drum *B* instead of drum *A*.

15.13 A mixing drum of 125-mm outside radius rests on two casters, each of 25-mm radius. The drum executes 12 rev during the time interval t, while its angular velocity is being increased uniformly from 25 to 45 rpm. Knowing that no slipping occurs between the drum and the casters, determine (a) the angular acceleration of the casters, (b) the time interval t.

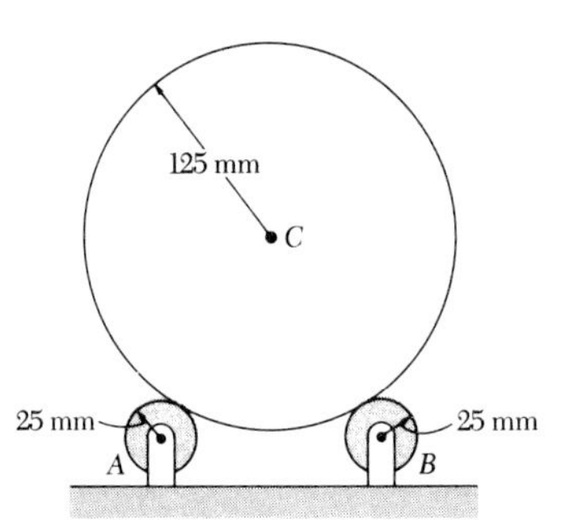

Fig. P15.13

15.14 A load is to be raised 20 ft by the hoisting system shown. Assuming gear A is initially at rest, accelerates uniformly to a speed of 120 rpm in 5 s, and then maintains a constant speed of 120 rpm, determine (a) the number of revolutions executed by gear A in raising the load, (b) the time required to raise the load.

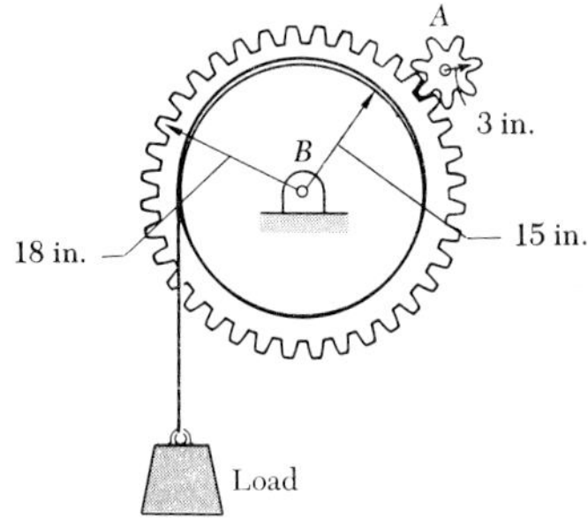

Fig. P15.14

15.15 A simple friction drive consists of two disks A and B. Initially, disk B has a clockwise angular velocity of 500 rpm, and disk A is at rest. It is known that disk B will coast to rest in 60 s. However, rather than waiting until both disks are at rest to bring them together, disk A is given a constant angular acceleration of 3 rad/s^2 counterclockwise. Determine (a) at what time the disks may be brought together if they are not to slip, (b) the angular velocity of each disk as contact is made.

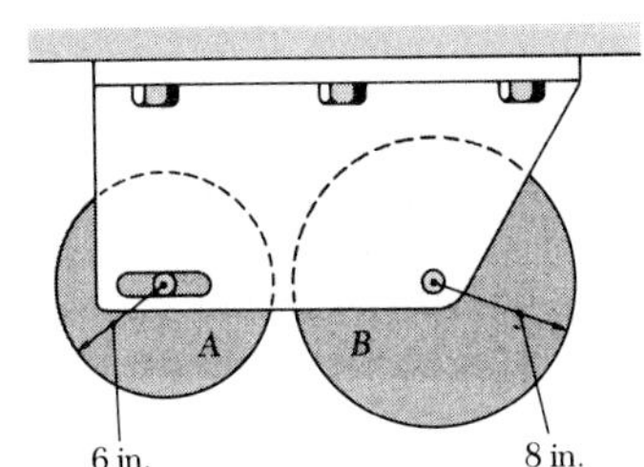

Fig. P15.15 and P15.16

15.16 Two friction wheels A and B are both rotating freely at 240 rpm counterclockwise when they are brought into contact. After 8 s of slippage, during which each wheel has a constant angular acceleration, wheel B reaches a final angular velocity of 60 rpm counterclockwise. Determine (a) the angular acceleration of each wheel during the period of slippage, (b) the time at which the angular velocity of wheel A is equal to zero.

SECTIONS 15.5 to 15.6

15.17 Collar B moves upward with a constant velocity of 1.8 m/s. At the instant when $\theta = 50°$, determine (a) the angular velocity of rod AB, (b) the velocity of end A of the rod.

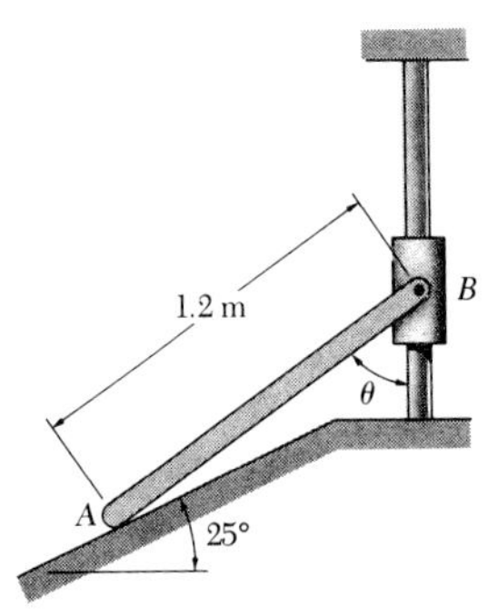

Fig. P15.17

15.18 Rod AB is 30 in. long and slides with its ends in contact with the floor and the inclined plane. End A moves with a constant velocity of 25 in./s to the right. At the instant when $\theta = 25°$, determine (a) the angular velocity of the rod, (b) the velocity of end B.

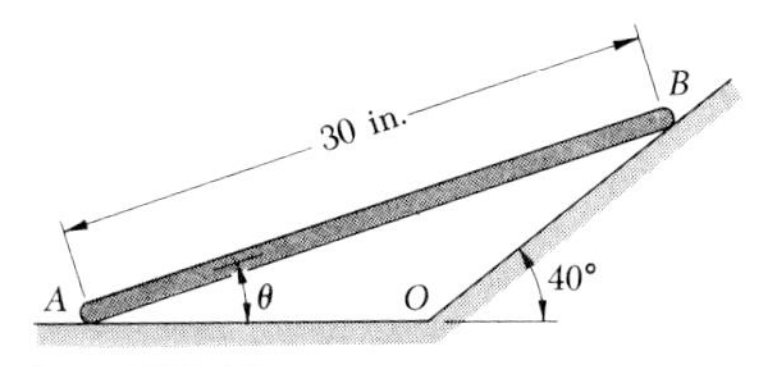

Fig. P15.18

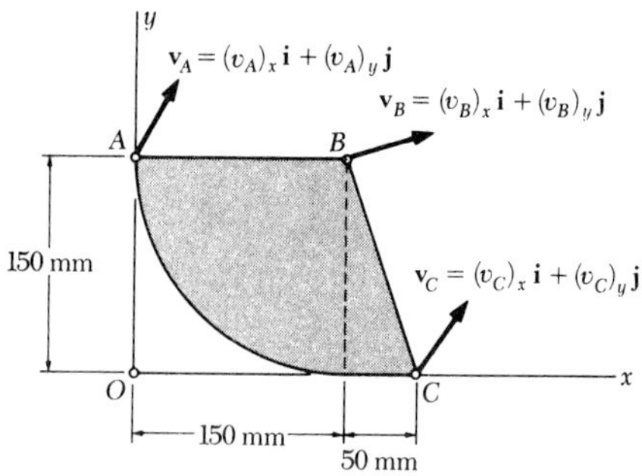

Fig. P15.19

15.19 The plate shown moves in the xy plane. Knowing that $(v_A)_x = 300$ mm/s, $(v_B)_y = -180$ mm/s, and $(v_C)_x = -150$ mm/s, determine (*a*) the angular velocity of the plate, (*b*) the velocity of point A.

15.20 The plate shown moves in the xy plane. Knowing that $(v_A)_x = 24$ in./s, $(v_B)_x = 8$ in./s, and $(v_C)_y = -20$ in./s, determine (*a*) the angular velocity of the plate, (*b*) the velocity of point B.

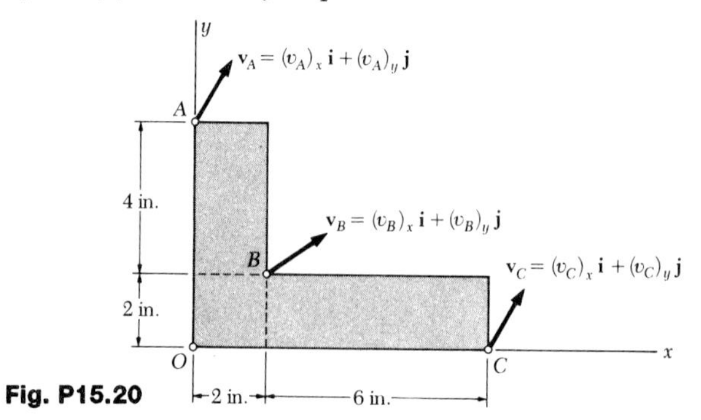

Fig. P15.20

15.21 In Prob. 15.20, determine (*a*) the velocity of point A, (*b*) the point of the plate with zero velocity.

15.22 In Prob. 15.19, determine (*a*) the velocity of point B, (*b*) the point of the plate with zero velocity.

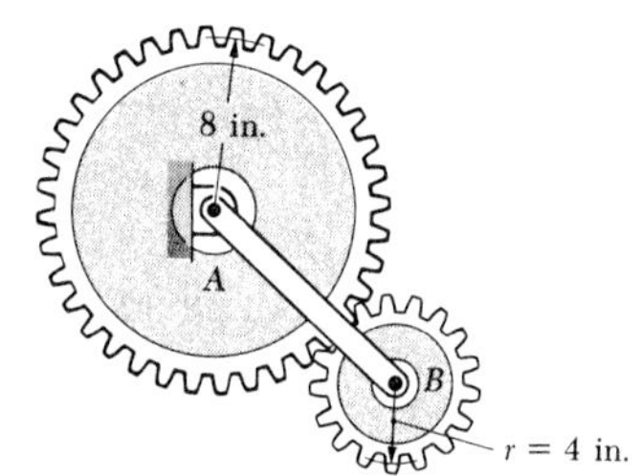

Fig. P15.23 and P15.24

15.23 Gear A rotates clockwise with a constant angular velocity of 60 rpm. Determine the angular velocity of gear B if the angular velocity of arm AB is (*a*) 40 rpm counterclockwise, (*b*) 40 rpm clockwise.

15.24 Gear A rotates clockwise with a constant angular velocity of 150 rpm. Knowing that at the same time arm AB rotates clockwise with a constant angular velocity of 125 rpm, determine the angular velocity of gear B.

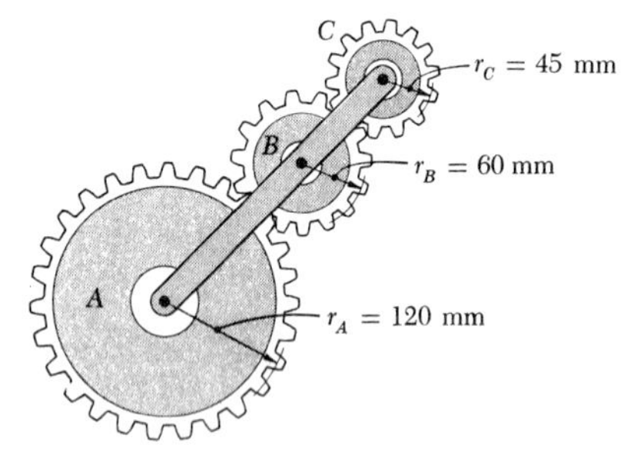

Fig. P15.25 and P15.26

15.25 Three gears A, B, and C are pinned at their centers to rod ABC. Knowing that gear A does not rotate, determine the angular velocity of gears B and C when the rod ABC rotates clockwise with a constant angular velocity of 48 rpm.

15.26 Three gears A, B, and C are pinned at their centers to rod ABC. Knowing that gear C does not rotate, determine the angular velocity of gears A and B when the rod ABC rotates clockwise with an angular velocity of 48 rpm.

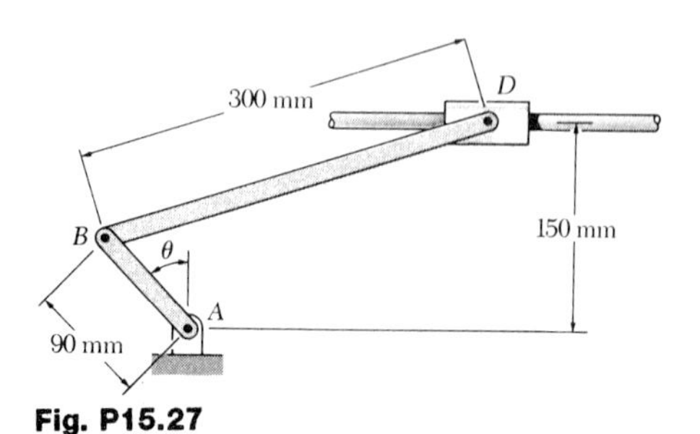

Fig. P15.27

15.27 Crank AB has a constant angular velocity of 200 rpm counterclockwise. Determine the angular velocity of rod BD and the velocity of collar D when (*a*) $\theta = 0$, (*b*) $\theta = 90°$, (*c*) $\theta = 180°$.

15.28 through 15.31 In the position shown, bar AB has a constant angular velocity of 3 rad/s counterclockwise. Determine the angular velocity of bars BD and DE.

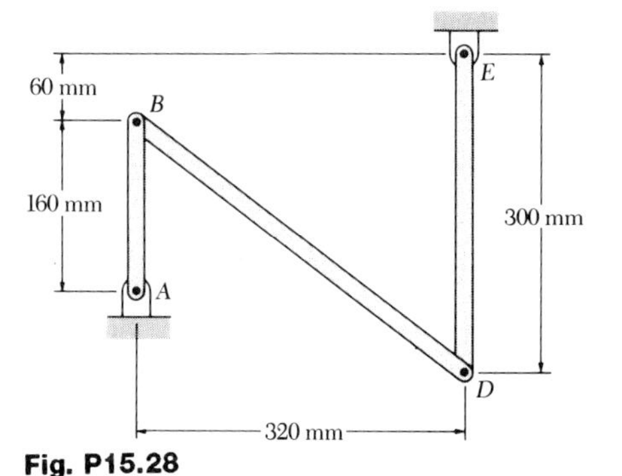

Fig. P15.28

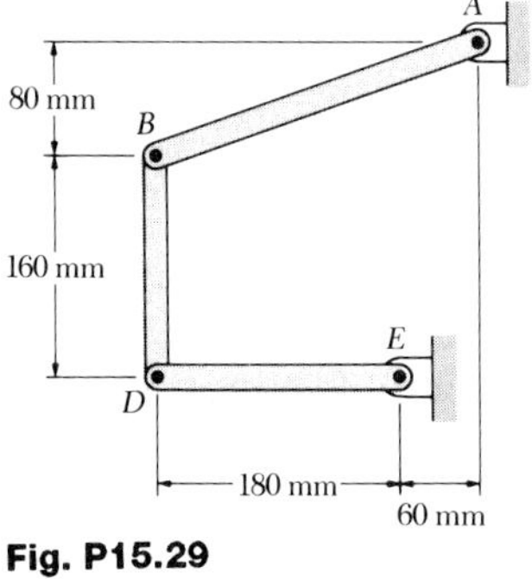
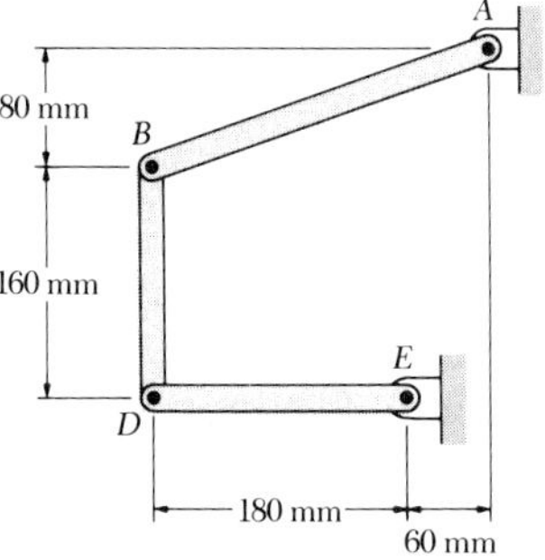

Fig. P15.29

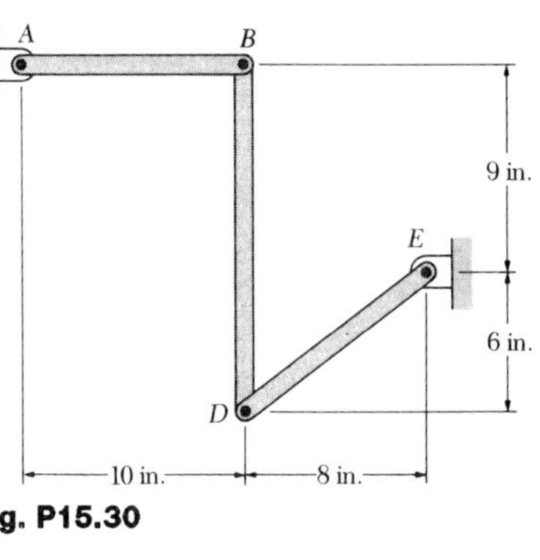

Fig. P15.30

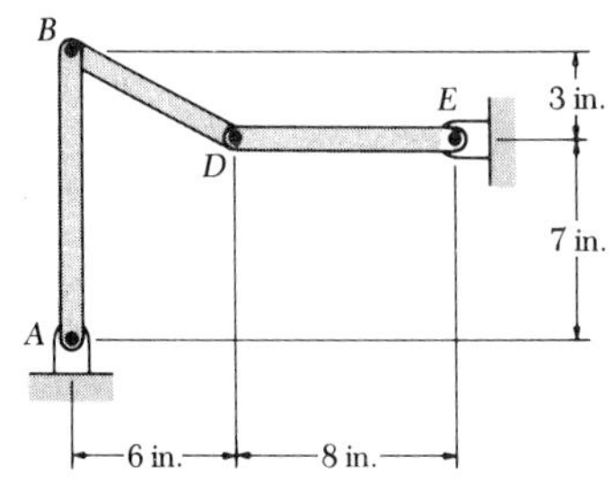

Fig. P15.31

15.32 For the gearing system shown, derive an expression for the angular velocity ω_C of gear C and show that ω_C is independent of the radius of gear B. Assume that point A is fixed and denote the angular velocities of rod ABC and gear A by ω_{ABC} and ω_A, respectively.

r_C = 45 mm

r_B = 60 mm

r_A = 120 mm

Fig. P15.32

SECTION 15.7

15.33 A helicopter moves horizontally in the x direction at a speed of 105 mi/h. Knowing that the main blades rotate clockwise at an angular velocity of 160 rpm, determine the instantaneous axis of rotation of the main blades.

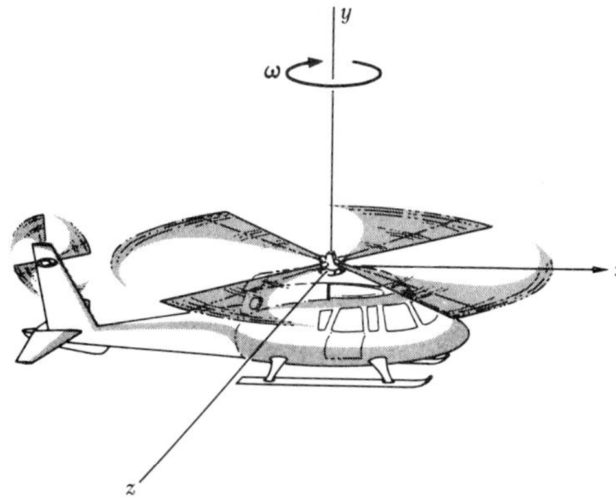

Fig. P15.33

15.34 The trolley shown moves to the left along a horizontal pipe support at a speed of 24 in./s. Knowing that the 5-in.-radius disk has an angular velocity of 8 rad/s counterclockwise, determine (a) the instantaneous center of rotation of the disk, (b) the velocity of point E.

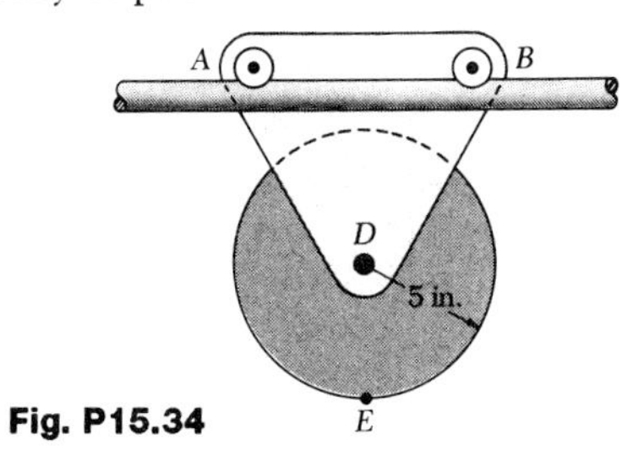

Fig. P15.34

15.35 A drum of radius 90 mm is mounted on a cylinder of radius 120 mm. A cord is wound around the drum, and its extremity D is pulled to the left at a constant velocity of 150 mm/s, causing the cylinder to roll without sliding. Determine (a) the angular velocity of the cylinder, (b) the velocity of the center of the cylinder, (c) the length of cord which is wound or unwound per second.

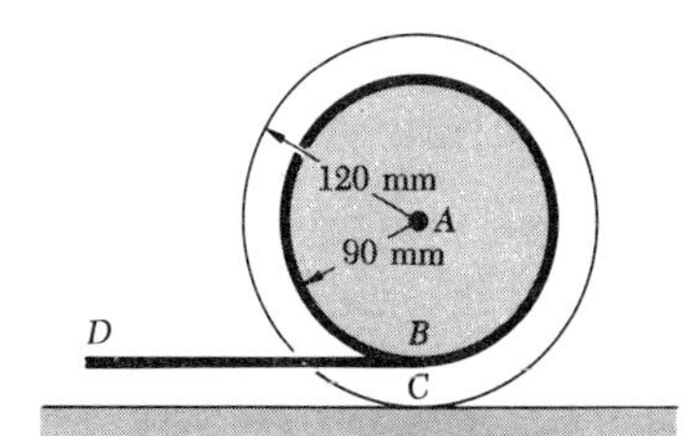

Fig. P15.35

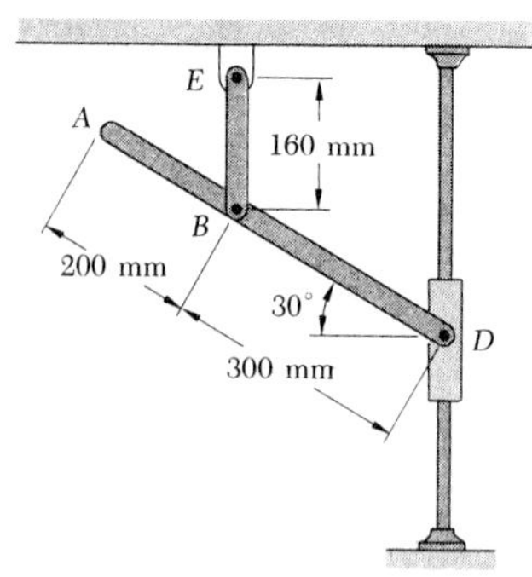

Fig. P15.36 and P15.37

15.36 Knowing that at the instant shown the angular velocity of rod BE is 3 rad/s counterclockwise, determine (*a*) the angular velocity of rod AD, (*b*) the velocity of collar D, (*c*) the velocity of point A.

15.37 Knowing that at the instant shown the velocity of collar D is 1.8 m/s upward, determine (*a*) the angular velocity of rod AD, (*b*) the velocity of point B, (*c*) the velocity of point A.

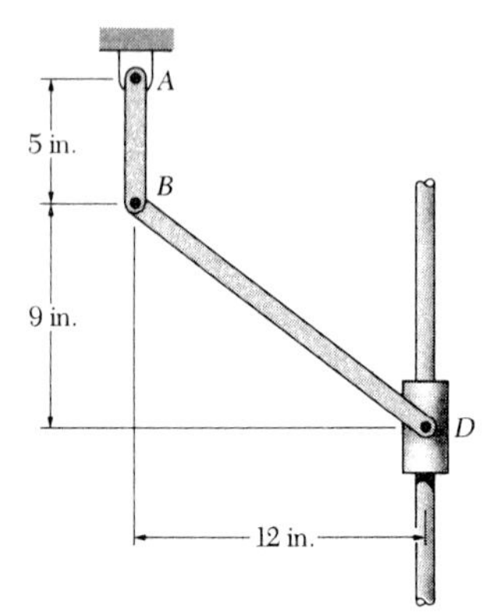

Fig. P15.38 and P15.39

15.38 Knowing that at the instant shown the velocity of collar D is 48 in./s upward, determine (*a*) the instantaneous center of rotation of link BD, (*b*) the angular velocities of crank AB and link BD, (*c*) the velocity of the midpoint of link BD.

15.39 Knowing that at the instant shown the angular velocity of crank AB is 2.7 rad/s clockwise, determine (*a*) the angular velocity of link BD, (*b*) the velocity of collar D, (*c*) the velocity of the midpoint of link BD.

15.40 The rod BDE is partially guided by a wheel at D which rolls in a vertical track. Knowing that at the instant shown the angular velocity of crank AB is 5 rad/s clockwise and that $\beta = 30°$, determine (*a*) the angular velocity of the rod, (*b*) the velocity of point E.

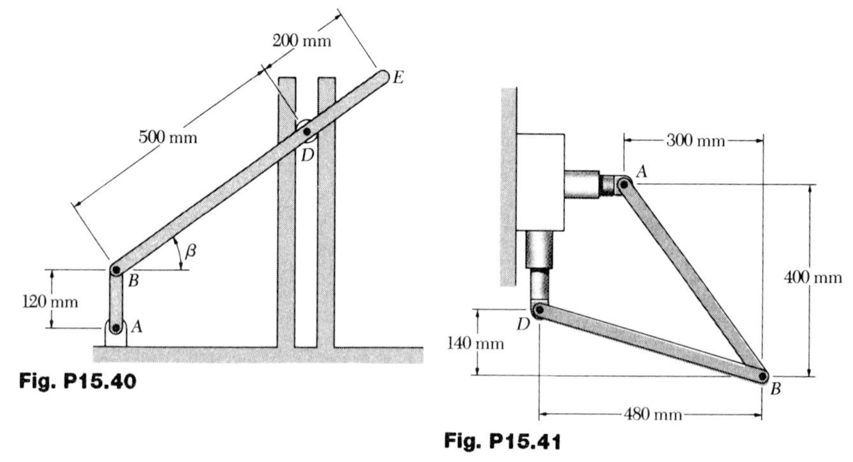

Fig. P15.40

Fig. P15.41

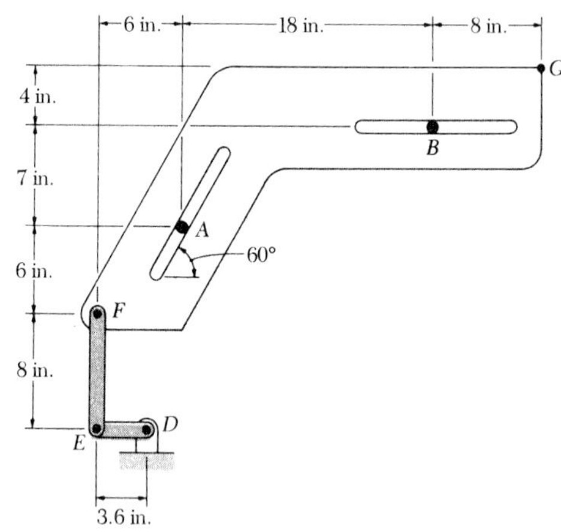

Fig. P15.43

15.41 Two links AB and BD, each 500 mm long, are connected at B and guided by hydraulic cylinders attached at A and D. Knowing that D is stationary and that the velocity of A is 1.5 m/s to the right, determine at the instant shown (*a*) the angular velocity of each link, (*b*) the velocity of B.

15.42 Solve Prob. 15.41, assuming that A is stationary and that the velocity of D is 1.5 m/s downward.

15.43 Two slots have been cut in the plate FG and the plate has been placed so that the slots fit two fixed pins A and B. Knowing that at the instant shown the angular velocity of crank DE is 8 rad/s clockwise, determine (*a*) the velocity of point F, (*b*) the velocity of point G.

15.44 In Prob. 15.43, determine the velocity of the point of the plate which is in contact with (*a*) pin A, (*b*) pin B.

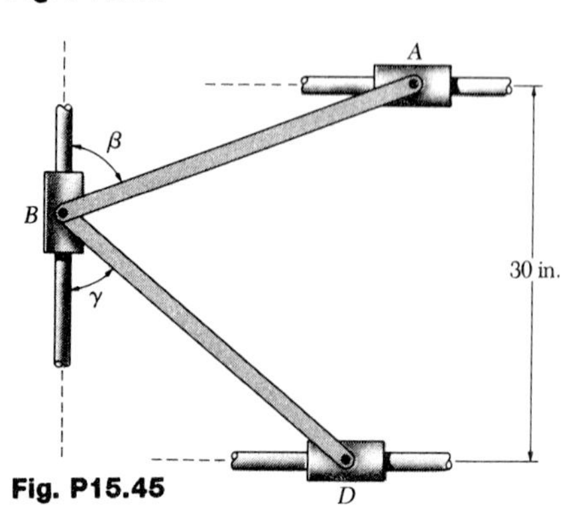

Fig. P15.45

15.45 Two links AB and BD, each of length 30 in., are connected to three collars. At the instant shown $\beta = 70°$ and the velocity of collar A is 90 in./s to the right. Determine (*a*) the corresponding value of γ, (*b*) the velocity of collar D.

15.46 Solve Prob. 15.45, assuming that $\beta = 55°$.

15.47 and 15.48 Two 500-mm rods are pin-connected at D as shown. Knowing that B moves to the right with a constant velocity of 480 mm/s, determine at the instant shown (a) the angular velocity of each rod, (b) the velocity of E.

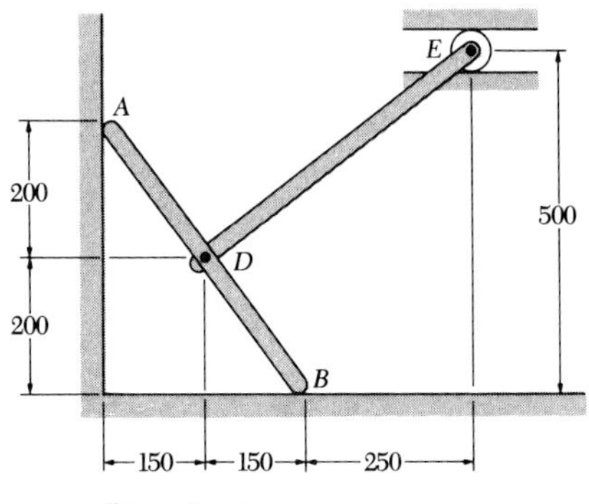

Fig. P15.47

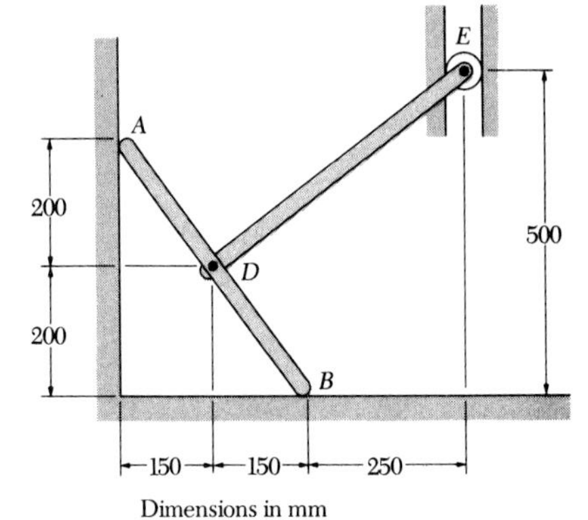

Fig. P15.48

SECTIONS 15.8 to 15.9

15.49 A tape is wrapped around a 10-in.-diameter disk which is at rest on a horizontal table. A force **P** applied as shown produces the following accelerations: $\mathbf{a}_A = 30$ in./s^2 to the right, $\boldsymbol{\alpha} = 4$ rad/s^2 counterclockwise as viewed from above. Determine the acceleration (a) of point B, (b) of point C.

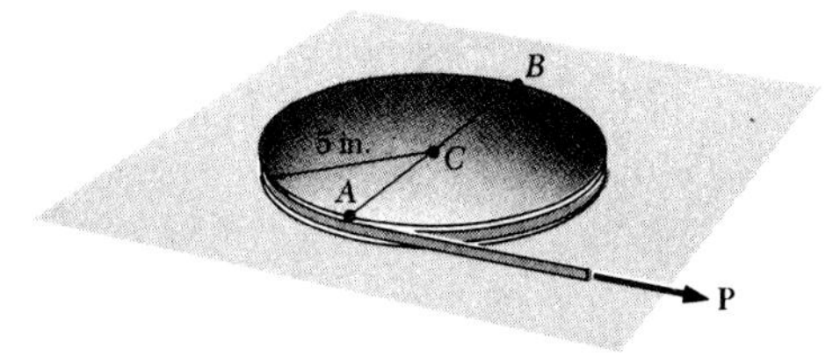

Fig. P15.49

15.50 In Prob. 15.49, determine the point of the plate which (a) has no acceleration, (b) has an acceleration of 22 in./s^2 to the right.

15.51 At the instant shown the center B of the double pulley has a velocity of 0.6 m/s and an acceleration of 2.4 m/s^2, both directed downward. Knowing that the cord wrapped around the inner pulley is attached to a fixed support at A, determine the acceleration of point D.

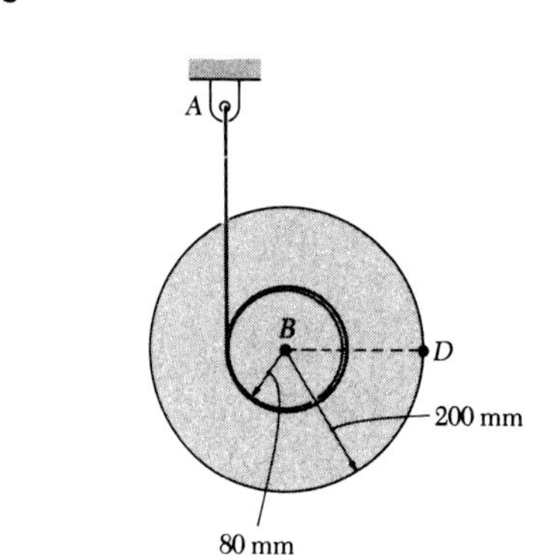

Fig. P15.51

15.52 An automobile is started from rest and has a constant acceleration of 7 ft/s^2 to the right. Knowing that the wheel shown rolls without sliding, determine the speed of the automobile at which the magnitude of the acceleration of point B of the wheel is 50 ft/s^2.

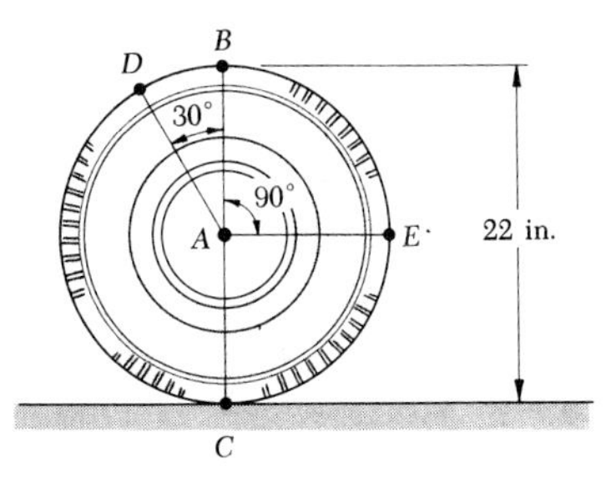

Fig. P15.52

15.53 In the two-cylinder air compressor shown the connecting rods BD and BE are each 8 in. long and the crank AB rotates about the fixed point A with a constant angular velocity of 1800 rpm clockwise. Determine the acceleration of each piston when $\theta = 0°$.

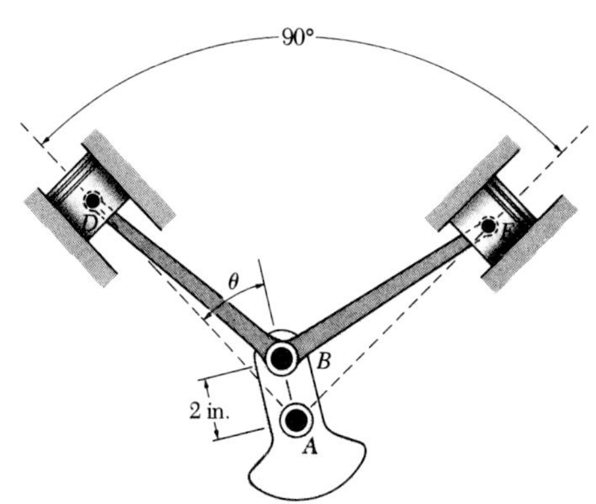

Fig. P15.53

15.54 In Prob. 15.53, determine the acceleration of piston E when $\theta = 45°$.

15.55 Arm AB rotates with a constant angular velocity of 90 rpm clockwise. Knowing that gear A does not rotate, determine the acceleration of the tooth of gear B which is in contact with gear A.

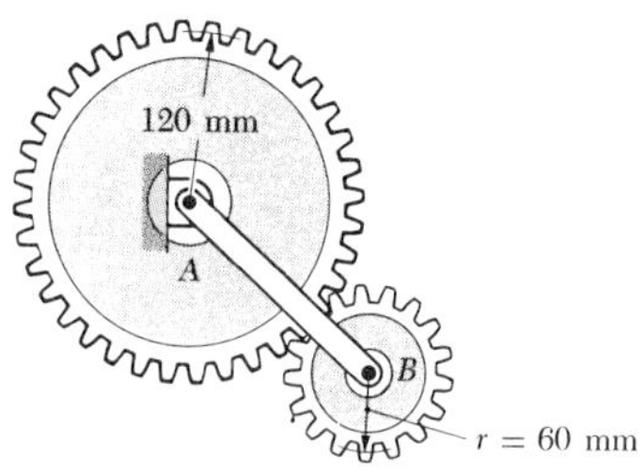

Fig. P15.55

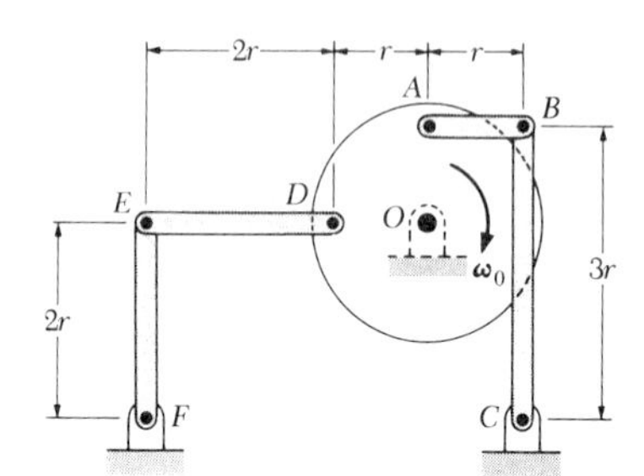

Fig. P15.56 and P15.57

15.56 At the instant shown, the disk rotates with a constant angular velocity ω_0 clockwise. Determine the angular velocities and the angular accelerations of the rods AB and BC.

15.57 At the instant shown, the disk rotates with a constant angular velocity ω_0 clockwise. Determine the angular velocities and the angular accelerations of the rods DE and EF.

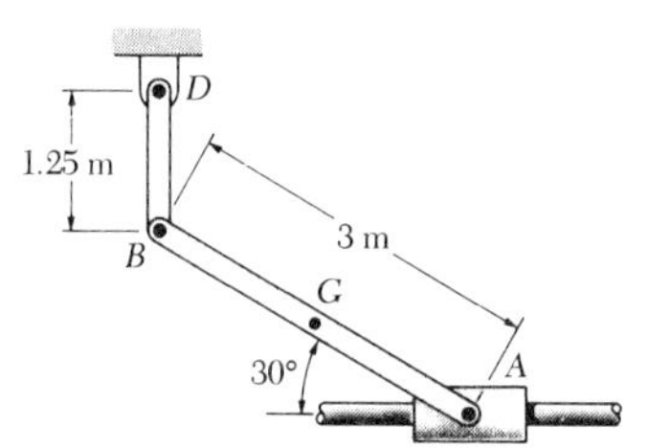

Fig. P15.58 and P15.59

15.58 End A of rod AB moves to the right with a constant velocity of 2.5 m/s. For the position shown, determine (a) the angular acceleration of rod AB, (b) the acceleration of the midpoint G of rod AB.

15.59 In the position shown, end A of rod AB has a velocity of 2.5 m/s and an acceleration of 1.5 m/s^2, both directed to the right. Determine (a) the angular acceleration of rod AB, (b) the acceleration of the midpoint G of rod AB.

15.60 In the position shown, end A of rod AB has a velocity of 3 ft/s and an acceleration of 2.5 ft/s^2, both directed to the left. Determine (a) the angular acceleration of rod AB, (b) the acceleration of the midpoint G of rod AB.

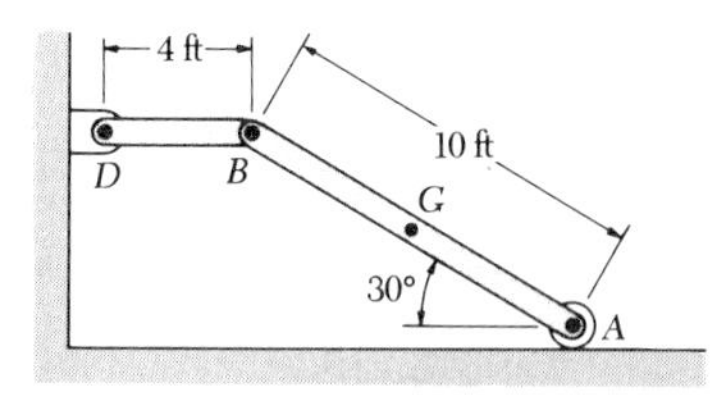

Fig. P15.60 and P15.61

15.61 End A of rod AB moves to the left with a constant velocity of 3 ft/s. For the position shown, determine (a) the angular acceleration of rod AB, (b) the acceleration of the midpoint G of rod AB.

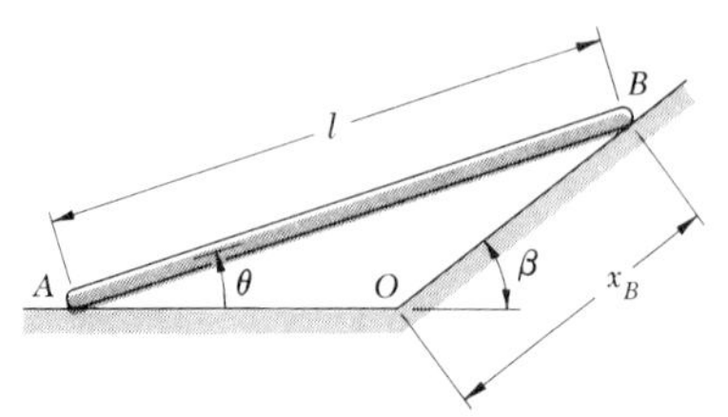

Fig. P15.62 and P15.63

15.62 Rod AB slides with its ends in contact with the floor and the inclined plane. Using the method of Sec. 15.9, derive an expression for the angular velocity of the rod in terms of v_B, θ, l, and β.

15.63 Derive an expression for the angular acceleration of the rod AB in terms of v_B, θ, l, and β, knowing that the acceleration of point B is zero.

15.64 The drive disk of the Scotch crosshead mechanism shown has an angular velocity $\boldsymbol{\omega}$ and an angular acceleration $\boldsymbol{\alpha}$, both directed clockwise. Using the method of Sec. 15.9, derive an expression for (*a*) the velocity of point *A*, (*b*) the acceleration of point *A*.

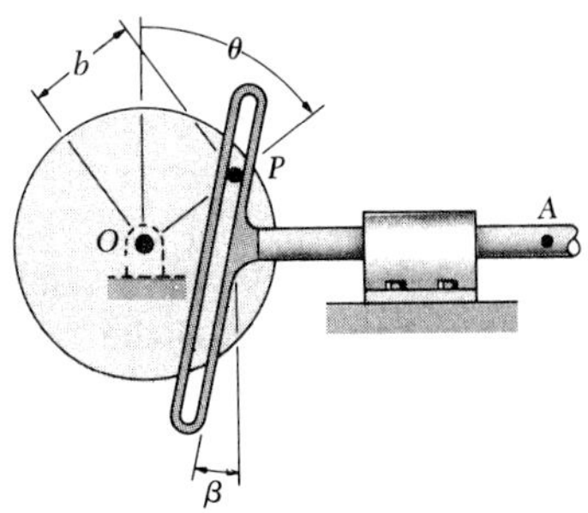

Fig. P15.64

15.65 The crank *AB* rotates with a constant clockwise angular velocity ω, and $\theta = 0$ at $t = 0$. Using the method of Sec. 15.9, derive an expression for the velocity of the piston *P* in terms of the time *t*.

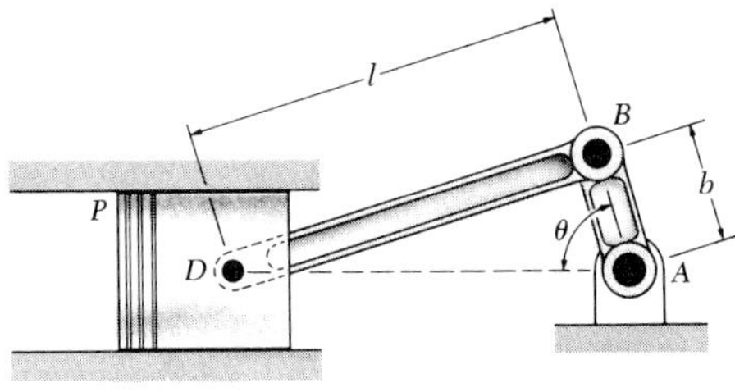

Fig. P15.65

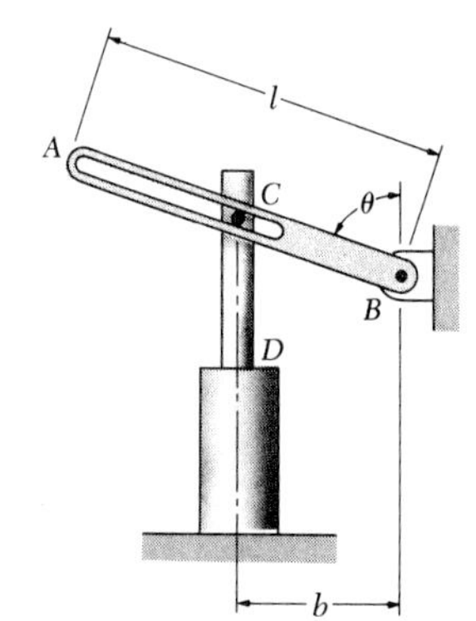

Fig. P15.66

15.66 Pin *C* is attached to rod *CD* and slides in a slot cut in arm *AB*. Knowing that rod *CD* moves vertically upward with a constant velocity $\mathbf{v}_0$, derive an expression for (*a*) the angular velocity of arm *AB*, (*b*) the components of the velocity of point *A*.

SECTIONS 15.10 to 15.11

15.67 and 15.68 Two rotating rods are connected by a slider block *P*. The rod attached at *A* rotates with a constant clockwise angular velocity $\boldsymbol{\omega}_A$. For the given data, determine for the position shown (*a*) the angular velocity of the rod attached at *B*, (*b*) the relative velocity of the slider block *P* with respect to the rod on which it slides.

15.67 $b = 8$ in., $\omega_A = 6$ rad/s.
15.68 $b = 250$ mm, $\omega_A = 8$ rad/s.

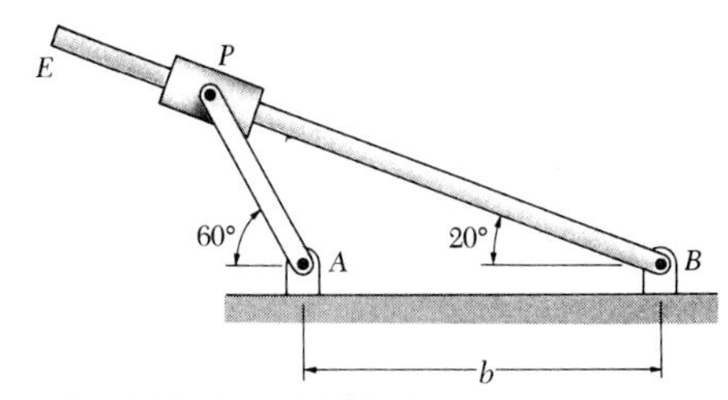

Fig. P15.67 and P15.69

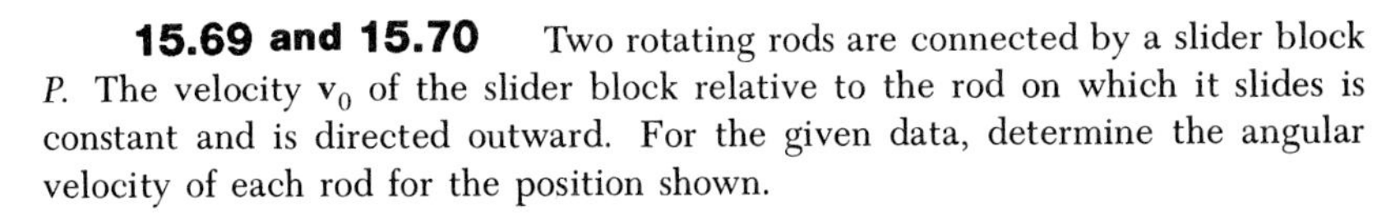

15.69 and 15.70 Two rotating rods are connected by a slider block *P*. The velocity $\mathbf{v}_0$ of the slider block relative to the rod on which it slides is constant and is directed outward. For the given data, determine the angular velocity of each rod for the position shown.

15.69 $b = 250$ mm, $v_0 = 360$ mm/s.
15.70 $b = 8$ in., $v_0 = 12$ in./s.

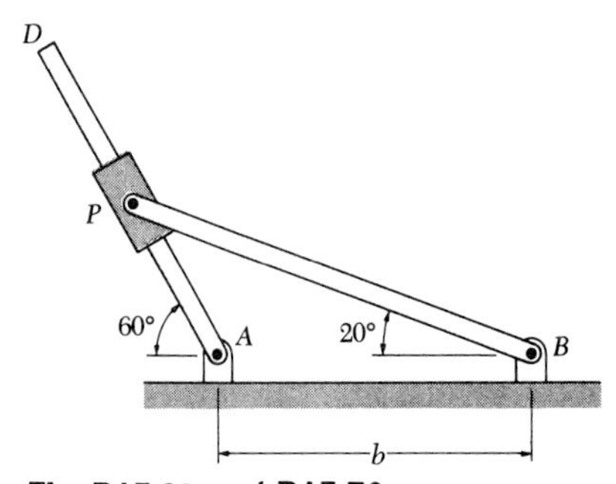

Fig. P15.68 and P15.70

15.71 Water flows at a constant rate through a straight pipe *OB* which rotates counterclockwise with a constant angular velocity of 150 rpm. If the velocity of the water relative to the pipe is 6 m/s, determine the total acceleration (*a*) of the particle of water P_1, (*b*) of the particle of water P_2.

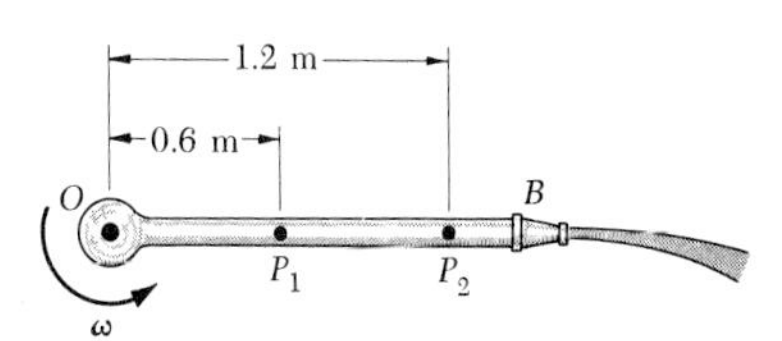

Fig. P15.71

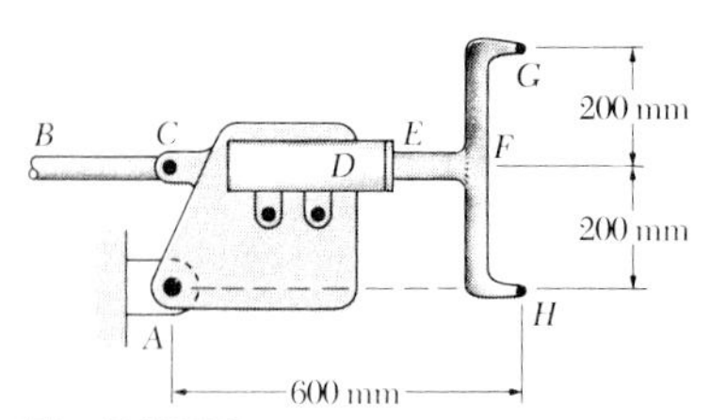

Fig. P15.72

15.72 In the automated welding setup shown, the position of the two welding tips G and H is controlled by the hydraulic cylinder D and the rod BC. The cylinder is bolted to the vertical plate which at the instant shown rotates counterclockwise about A with a constant angular velocity of 1.2 rad/s. Knowing that at the same instant the length EF of the welding asssembly is increasing at the constant rate of 300 mm/s, determine (*a*) the velocity of tip H, (*b*) the acceleration of tip H.

15.73 In Prob. 15.72, determine (*a*) the velocity of tip G, (*b*) the acceleration of tip G.

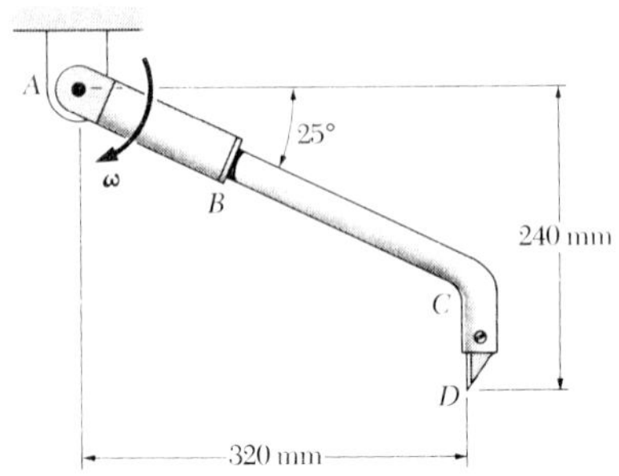

Fig. P15.74

15.74 The motion of blade D is controlled by the robot arm ABC. At the instant shown the arm is rotating clockwise at the constant rate $\omega = 1.5$ rad/s and the length of portion BC of the arm is being decreased at the constant rate of 180 mm/s. Determine (*a*) the velocity of D, (*b*) the acceleration of D.

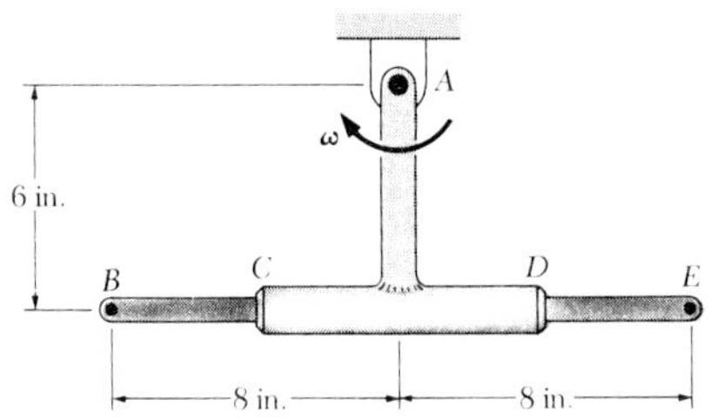

Fig. P15.75

15.75 The hydraulic cylinder CD is welded to an arm which rotates clockwise about A at the constant rate $\omega = 3$ rad/s. Knowing that in the position shown the rod BE is being moved to the right at the constant rate of 12 in./s with respect to the cylinder, determine (*a*) the velocity of point B, (*b*) the acceleration of point B.

15.76 In Prob. 15.75, determine (*a*) the velocity of point E, (*b*) the acceleration of point E.

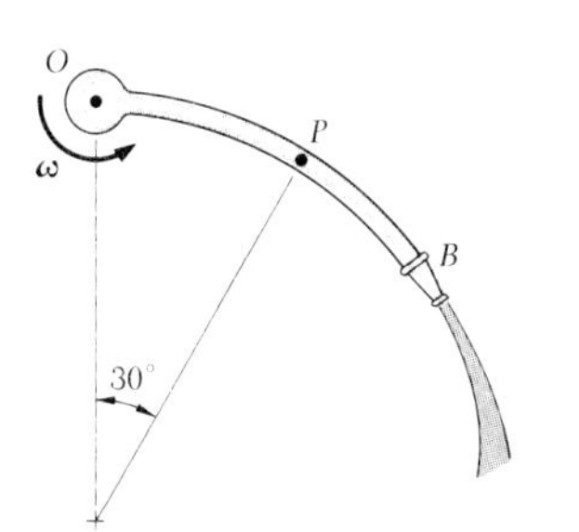

Fig. P15.77

15.77 Water flows through the curved pipe OB, which has a uniform radius of 15 in. and which rotates with a constant counterclockwise angular velocity of 120 rpm. If the velocity of the water relative to the pipe is 40 ft/s, determine the total acceleration of the particle of water P.

15.78 Solve Prob. 15.77, assuming that the curved pipe rotates with a constant clockwise angular velocity of 120 rpm.

15.79 Rod AD is bent in the shape of an arc of circle of radius $b = 120$ mm. The position of the rod is controlled by pin B which slides in a horizontal slot and also slides along the rod. Knowing that at the instant shown pin B moves to the right with a constant speed of 40 mm/s, determine (*a*) the angular velocity of the rod, (*b*) the angular acceleration of the rod.

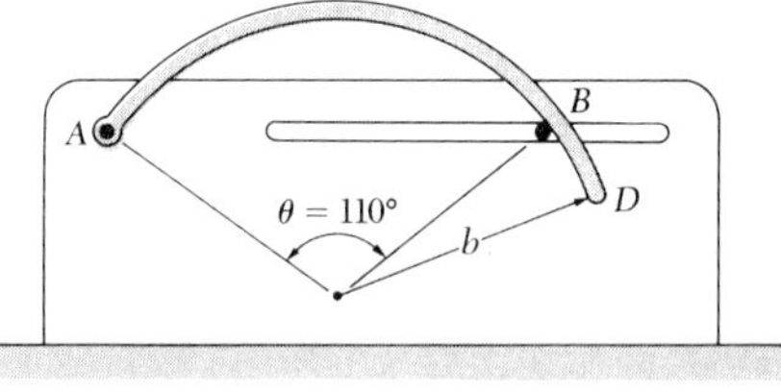

Fig. P15.79

15.80 Solve Prob. 15.79 when $\theta = 90°$.

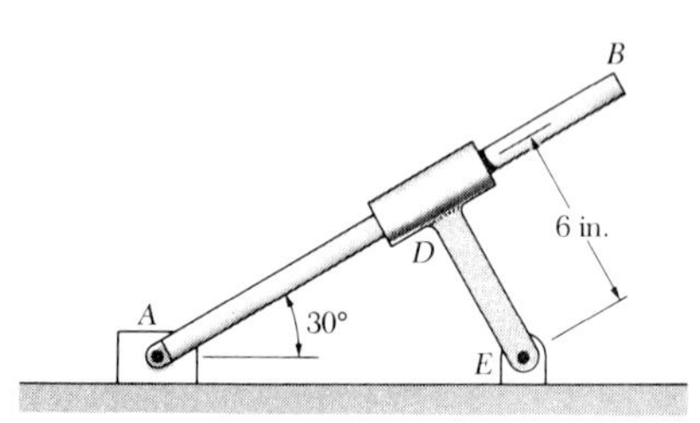

Fig. P15.81

15.81 Rod AB passes through a collar which is welded to link DE. Knowing that at the instant shown block A moves to the right with a constant speed of 75 in./s, determine (*a*) the angular velocity of rod AB, (*b*) the velocity relative to the collar of the point of the rod in contact with the collar, (*c*) the acceleration of the point of the rod in contact with the collar. (*Hint.* Rod AB and link DE have the same $\boldsymbol{\omega}$ and the same $\boldsymbol{\alpha}$.)

15.82 Solve Prob. 15.81, assuming that block A moves to the left with a constant speed of 75 in./s.

15.83 Plate ABD and rod OB are rigidly connected and rotate about the ball-and-socket joint O with an angular velocity $\boldsymbol{\omega} = \omega_x\mathbf{i} + \omega_y\mathbf{j} + \omega_z\mathbf{k}$. Knowing that $\mathbf{v}_A = (14 \text{ in./s})\mathbf{i} + (16 \text{ in./s})\mathbf{j} + (v_A)_z\mathbf{k}$ and $\omega_x = 2$ rad/s, determine (*a*) the angular velocity of the assembly, (*b*) the velocity of point D.

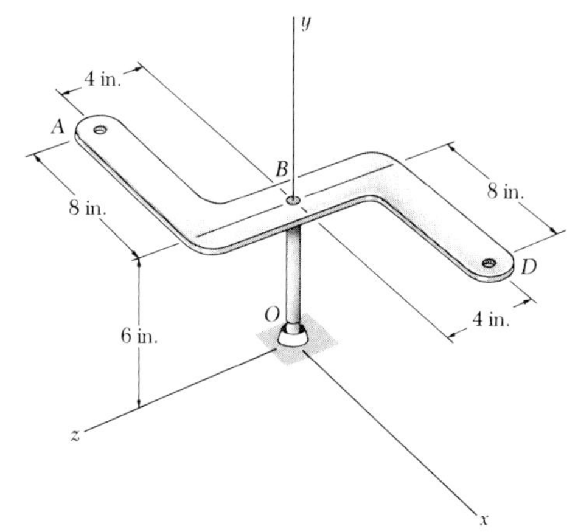

Fig. P15.83

15.84 The rigid body shown rotates about the origin of coordinates with an angular velocity $\boldsymbol{\omega}$. Denoting the velocity of point A by $\mathbf{v}_A = (v_A)_x\mathbf{i} + (v_A)_y\mathbf{j} + (v_A)_z\mathbf{k}$, and knowing that $(v_A)_x = 150$ mm/s and $(v_A)_y = -75$ mm/s, determine the velocity component $(v_A)_z$.

15.85 The rigid body shown rotates about the origin of coordinates with an angular velocity $\boldsymbol{\omega} = \omega_x\mathbf{i} + \omega_y\mathbf{j} + \omega_z\mathbf{k}$. Knowing that $(v_A)_y = 165$ mm/s, $(v_B)_y = 315$ mm/s, and $\omega_y = -3$ rad/s, determine (*a*) the angular velocity of the body, (*b*) the velocities of points A and B.

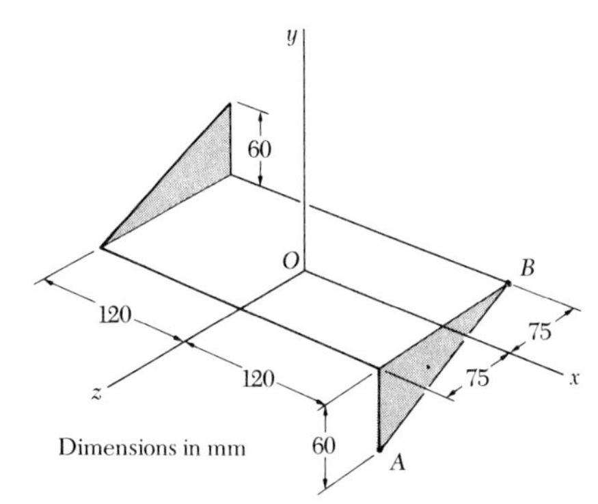

Fig. P15.84 and P15.85

15.86 The disk of a portable sander rotates at the constant rate $\omega_1 =$ 4400 rpm as shown. Determine the angular acceleration of the disk as a worker rotates the sander about the z axis with an angular velocity of 0.5 rad/s and an angular acceleration of 2.5 rad/s^2, both clockwise when viewed from the positive z axis.

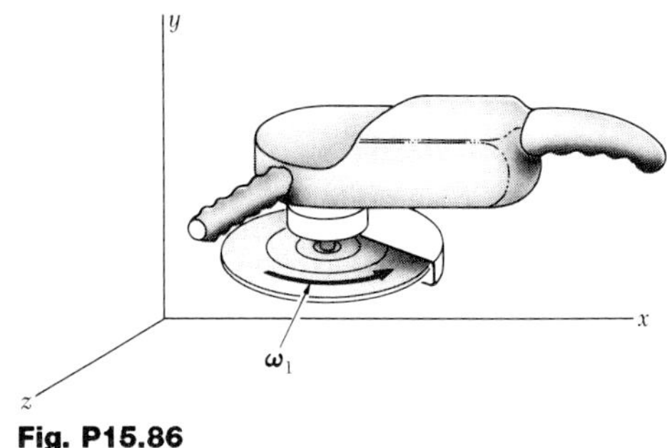

Fig. P15.86

15.87 The fan of an automobile engine rotates about a horizontal axis at the rate of 2500 rpm in a clockwise sense when viewed from the rear of the engine. Knowing that the automobile is turning right along a path of radius 40 ft at a constant speed of 9 mi/h, determine the angular acceleration of the fan at the instant the automobile is moving due north.

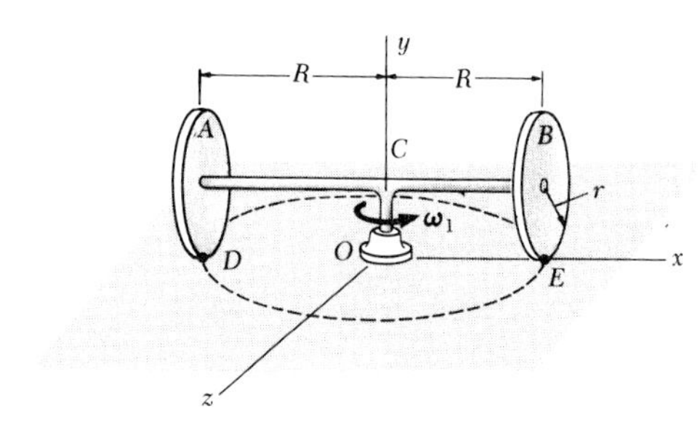

Fig. P15.88

15.88 Two disks A and B are mounted on an axle of length $2R$ and roll without sliding on a horizontal floor. Knowing that the axle rotates with a constant angular velocity $\boldsymbol{\omega}_1$, determine (*a*) the angular velocity of disk A, (*b*) the angular acceleration of disk A.

15.89 In Prob. 15.88, determine (*a*) the angular velocity of disk B, (*b*) the angular acceleration of disk B.

15.90 The robot component shown rotates with a constant angular velocity $\boldsymbol{\omega}_1$ of 2.5 rad/s about the x axis, while arm BC rotates about the z axis with an angular velocity $\boldsymbol{\omega}_2$ which, at the instant shown, has a magnitude $\omega_2 = 3$ rad/s and increases at the rate $\dot{\omega}_2 = 4$ rad/s^2. Determine the angular acceleration of arm BC.

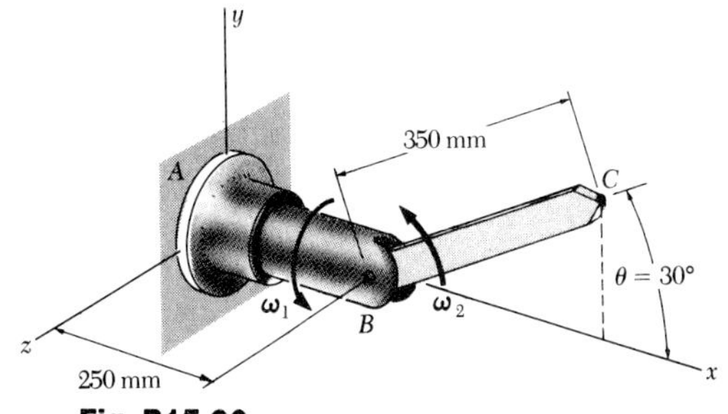

Fig. P15.90

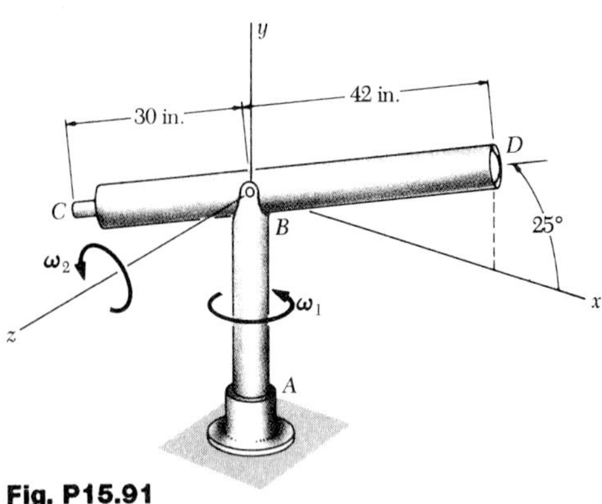

Fig. P15.91

15.91 At the instant shown the telescope CD is being rotated simultaneously at the constant rate $\omega_1 = 0.15$ rad/s about the y axis, and at the constant rate $\omega_2 = 0.25$ rad/s about the z axis. Determine the velocity and acceleration of the center D of the objective lens.

15.92 In Prob. 15.91, determine the velocity and acceleration of the eyepiece C.

15.93 A disk of radius r is mounted on an axle of length $2r$. The axle is attached to a vertical shaft AD which rotates at the constant rate ω_1 and the disk rotates about the axle AB at the constant rate ω_2. Knowing that the angle θ remains constant and that the rim of the disk touches the y axis, determine (*a*) the angular acceleration of the disk, (*b*) the velocity of point C of the disk, (*c*) the acceleration of point C of the disk.

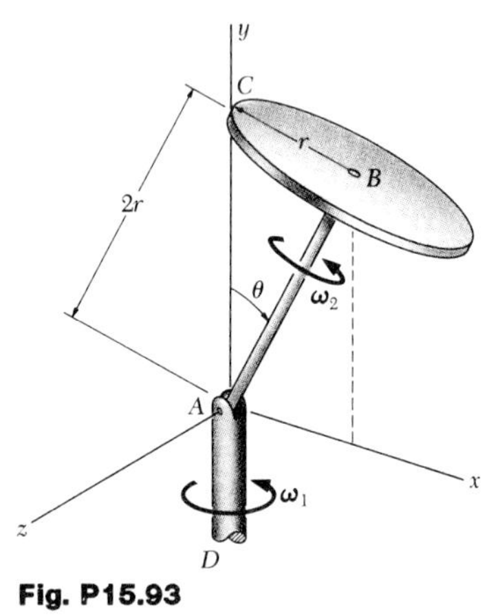

Fig. P15.93

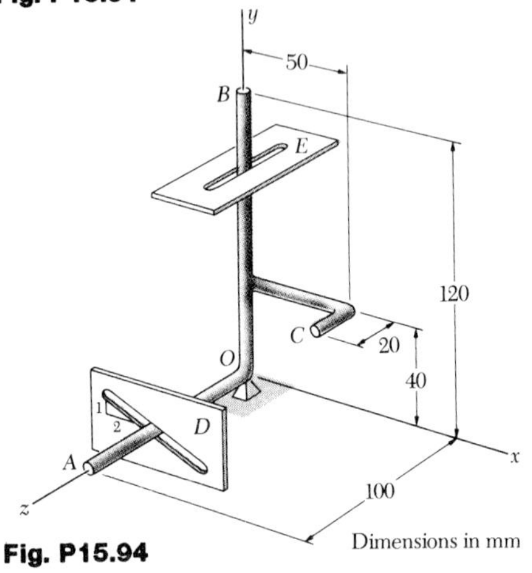

Fig. P15.94

15.94 Several rods are brazed together to form the robotic guide arm shown which is attached to a ball-and-socket joint at O. Rod OA slides in a straight inclined slot while rod OB slides in a slot parallel to the z axis. Knowing that at the instant shown $\mathbf{v}_B = (180 \text{ mm/s})\mathbf{k}$, determine (*a*) the angular velocity of the guide arm, (*b*) the velocity of point A, (*c*) the velocity of point C.

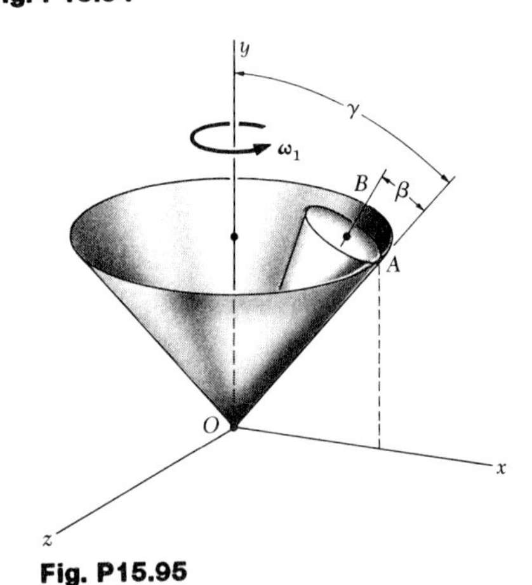

Fig. P15.95

15.95 The small cone shown rolls without slipping on the inside surface of the large fixed cone. Denoting by $\boldsymbol{\omega}_1$ the constant angular velocity of the axis OB about the y axis, determine in terms of ω_1, β, and γ, (*a*) the rate of spin of the cone about the axis OB, (*b*) the total angular velocity of the cone, (*c*) the angular acceleration of the cone.

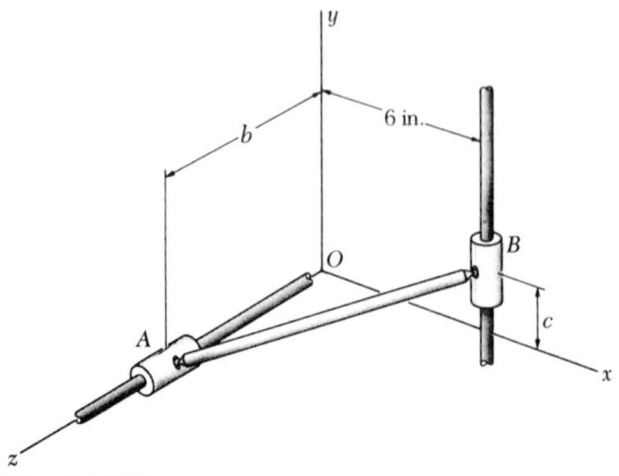

Fig. P15.96

15.96 Rod AB, of length 11 in., is connected by ball-and-socket joints to collars A and B, which slide along the two rods shown. Knowing that collar B moves downward with a constant speed of 54 in./s, determine the velocity of collar A when $c = 2$ in.

15.97 Rods *BC* and *BD* are each 630 mm long and are connected by ball-and-socket joints to collars which may slide on the fixed rods shown. Knowing that at the instant shown, collar *B* moves toward *A* with a constant speed of 260 mm/s, determine the velocity of collar *C*.

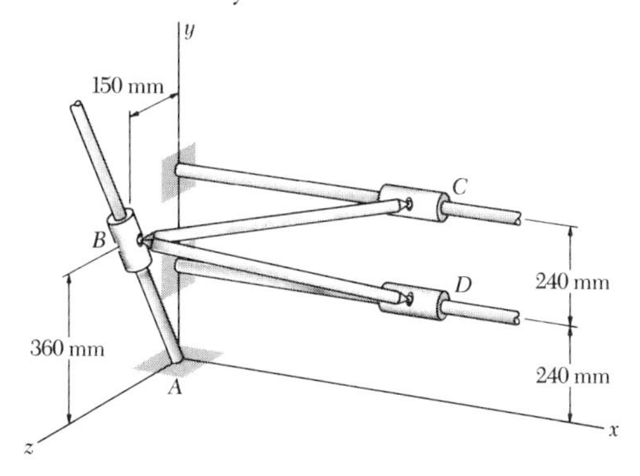

Fig. P15.97

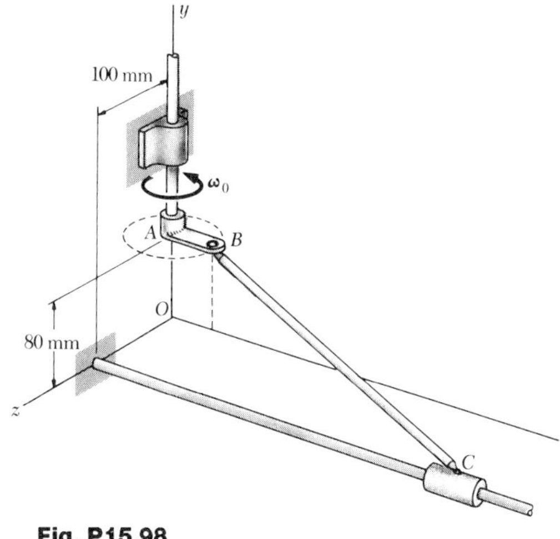

Fig. P15.98

15.98 Rod *BC*, of length 420 mm, is connected by ball-and-socket joints to the rotating arm *AB* and to the collar *C*. Arm *AB* is of length 60 mm and rotates in a horizontal plane at the constant rate $\omega_0 = 20$ rad/s. At the instant shown, when arm *AB* is parallel to the *x* axis, determine the velocity of collar *C*.

SECTIONS 15.14 to 15.15

15.99 The bent rod *ABC* rotates at the constant rate $\omega_1 = 6$ rad/s. Knowing that the collar *D* moves toward end *C* at a constant relative speed $u = 100$ in./s, determine for the position shown (*a*) the velocity of *D*, (*b*) the acceleration of *D*.

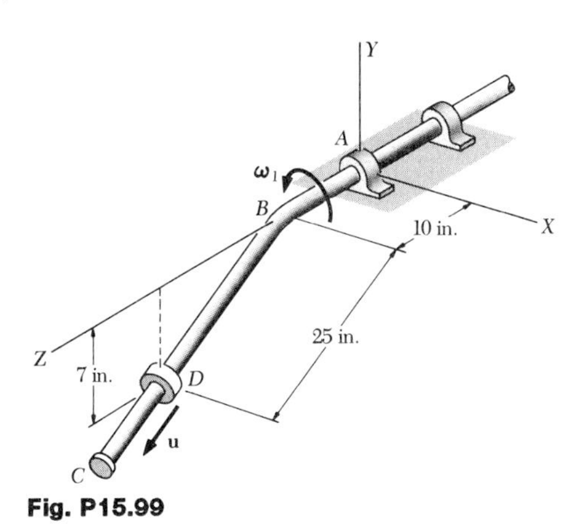

Fig. P15.99

15.100 The bent rod *ABC* rotates at the constant rate $\omega_1 = 4$ rad/s. Knowing that the collar *D* moves downward along the rod at a constant relative speed $u = 65$ in./s, determine for the position shown (*a*) the velocity of *D*, (*b*) the acceleration of *D*.

15.101 Solve Prob. 15.100, assuming that at the instant shown the angular velocity $\boldsymbol{\omega}_1$ of the rod is 4 rad/s and is decreasing at the rate of 10 rad/s^2, while the relative speed *u* of the collar is 65 in./s and is decreasing at the rate of 208 in./s^2.

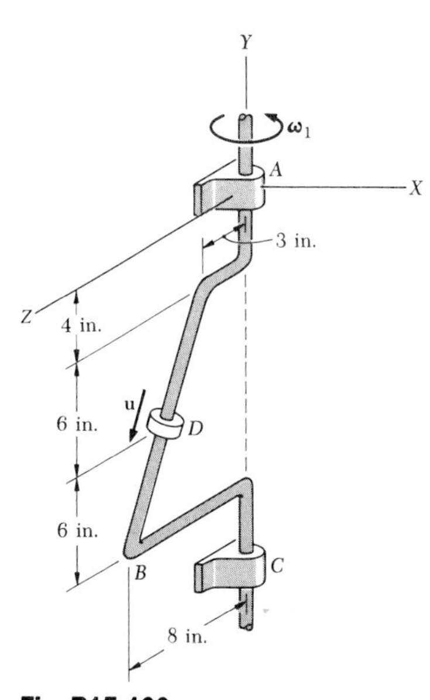

Fig. P15.100

15.102 Manufactured items are spray-painted as they pass through the automated work station shown. Knowing that the bent pipe ACE rotates at the constant rate $\omega_1 = 0.6$ rad/s and that at point D the paint moves through the pipe at a constant relative speed $u = 150$ mm/s, determine for the position shown (a) the velocity of the paint at D, (b) the acceleration of the paint at D.

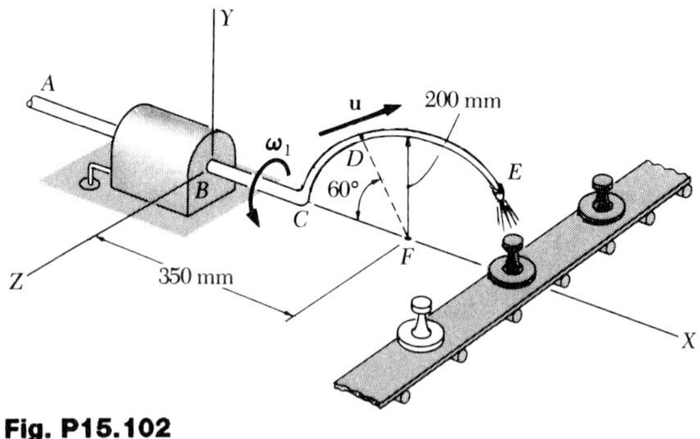

Fig. P15.102

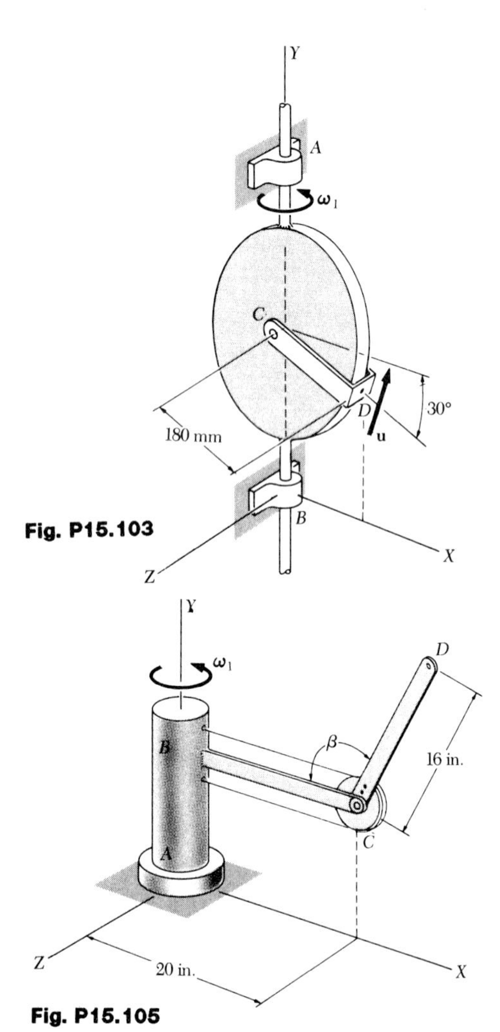

Fig. P15.103

15.103 The circular plate shown rotates about its vertical diameter at the constant rate $\omega_1 = 8$ rad/s. Knowing that in the position shown the disk lies in the XY plane and point D of strap CD moves upward at a constant relative speed $u = 1.2$ m/s, determine (a) the velocity of D, (b) the acceleration of D.

15.104 Solve Prob. 15.103, assuming that at the instant shown the angular velocity $\boldsymbol{\omega}_1$ of the plate is 8 rad/s and is decreasing at the rate of 25 rad/s^2 while the relative speed u of point D of strap CD is 1.2 m/s and is decreasing at the rate of 3 m/s^2.

Fig. P15.105

15.105 The body AB and rod BC of the robot component shown rotate at the constant rate $\omega_1 = 0.60$ rad/s about the Y axis. Simultaneously a wire-and-pulley control causes arm CD to rotate about C at the constant rate $\omega_2 = d\beta/dt = 0.45$ rad/s. Knowing that $\beta = 120°$, determine (a) the angular acceleration of arm CD, (b) the velocity of D, (c) the acceleration of D.

15.106 Solve Prob. 15.105, assuming that $\beta = 60°$.

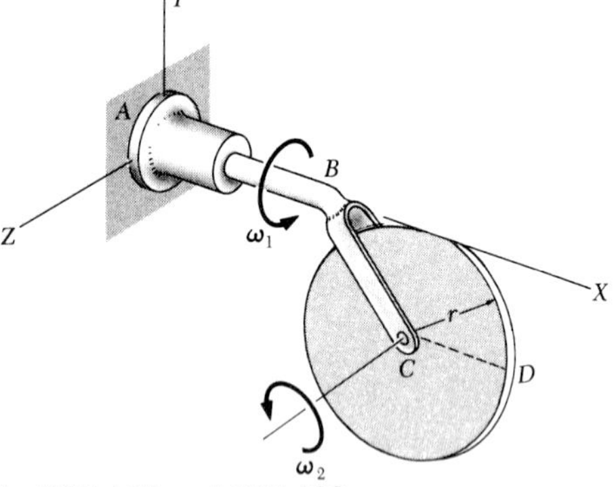

Fig. P15.107 and P15.108

15.107 A disk of radius r rotates at the constant rate ω_2 with respect to the arm ABC, which itself rotates at the constant rate ω_1 about the X axis. Determine (a) the angular velocity and angular acceleration of the disk, (b) the velocity and acceleration of point D on the rim of the disk.

15.108 A disk of radius $r = 120$ mm rotates at the constant rate $\omega_2 = 5$ rad/s with respect to the arm ABC, which itself rotates at the constant rate $\omega_1 = 4$ rad/s about the X axis. Determine the velocity and acceleration of point D on the rim of the disk.

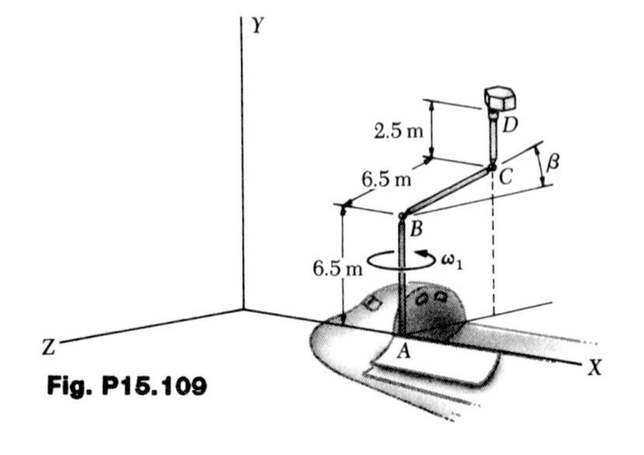

Fig. P15.109

15.109 The remote manipulator system (RMS) shown is used to deploy payloads from the cargo bay of space shuttles. At the instant shown, the whole RMS is rotating at the constant rate $\omega_1 = 0.02$ rad/s about the axis AB. At the same time, portion BCD rotates as a rigid body at the constant rate $\omega_2 = d\beta/dt = 0.03$ rad/s about an axis through B parallel to the X axis. Knowing that $\beta = 30°$, determine (a) the angular acceleration of BCD, (b) the velocity of D, (c) the acceleration of D.

15.110 In the portion of an industrial robot shown plate ABC rotates at the constant rate $\omega_1 = 1.8$ rad/s with respect to component DEG. At the same time, the entire unit rotates about the Y axis at the constant rate $\omega_2 = 1.5$ rad/s. Knowing that at the instant shown DE is parallel to the X axis and plate ABC is parallel to the YZ plane, determine (a) the velocity of point A, (b) the acceleration of point A.

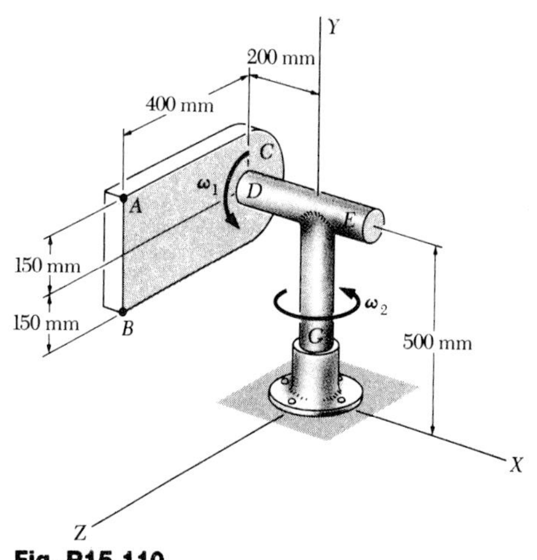

Fig. P15.110

15.111 The 120-ft blades of a wind-turbine generator rotate at the constant rate $\omega_1 = 25$ rpm. Knowing that at the instant shown the entire unit is being rotated about the Y axis at the constant rate $\omega_2 = 0.1$ rad/s, determine (a) the angular acceleration of the blades, (b) the velocity and acceleration of blade tip B.

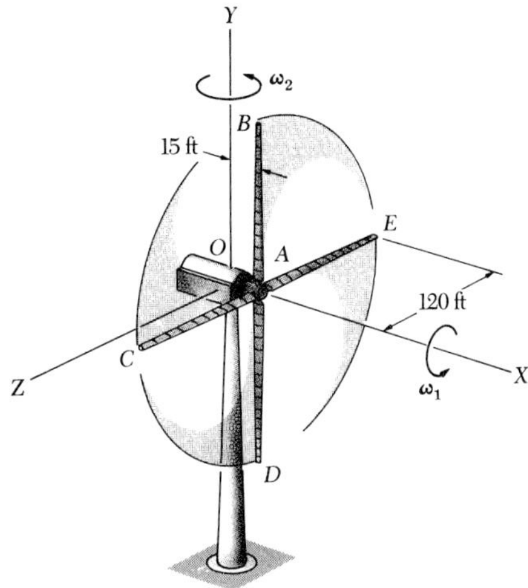

Fig. P15.111

15.112 In Prob. 15.111, determine the velocity and acceleration of (a) blade tip C, (b) blade tip D.

15.113 The position of the stylus tip A is controlled by the robot shown. In the position shown the stylus moves at a constant speed $u = 150$ mm/s relative to the solenoid BC. At the same time, arm CD rotates at the constant rate $\omega_2 = 1.6$ rad/s with respect to component DEG. Knowing that the entire robot rotates about the X axis at the constant rate $\omega_1 = 1.2$ rad/s, determine (a) the velocity of A, (b) the acceleration of A.

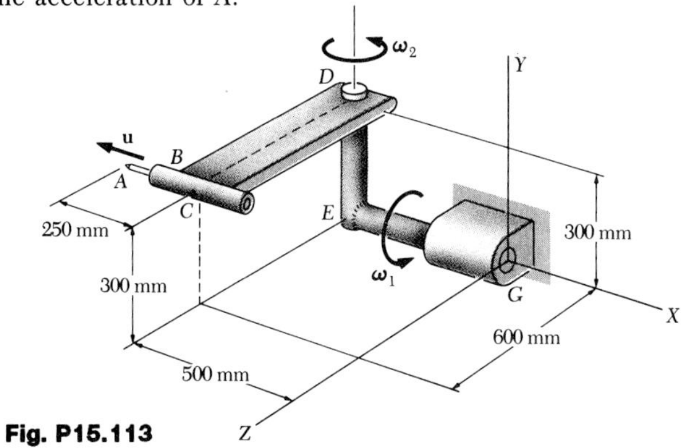

Fig. P15.113

15.114 A square plate of side $2r$ is welded to a vertical shaft which rotates with a constant angular velocity ω_1. At the same time, rod AB of length r rotates about the center of the plate with a constant angular velocity ω_2 with respect to the plate. For the position of the plate shown, determine the acceleration of end B of the rod if (a) $\theta = 0$, (b) $\theta = 90°$, (c) $\theta = 180°$.

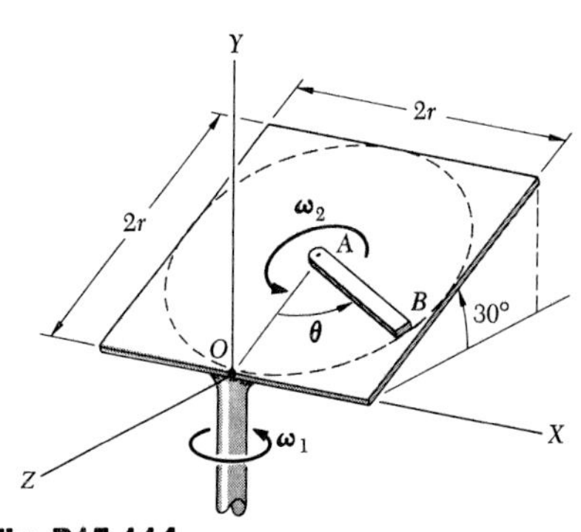

Fig. P15.114

CHAPTER 15 COMPUTER PROBLEMS

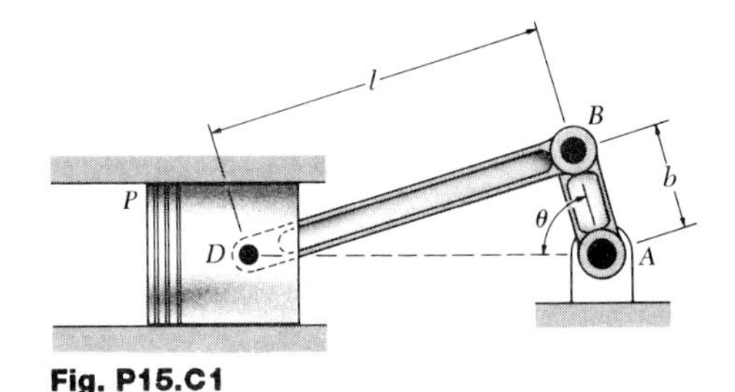

Fig. P15.C1

15.C1 In the engine system shown, $l = 10$ in. and $b = 3$ in.; the crank AB rotates with a constant angular velocity of 750 rpm clockwise. Derive expressions for the velocity of the piston and the angular velocity of the connecting rod as functions of time. Plot the velocity of the piston and the angular velocity of the connecting rod as functions of time for one revolution of the crank, setting $t = 0$ when $\theta = 0$.

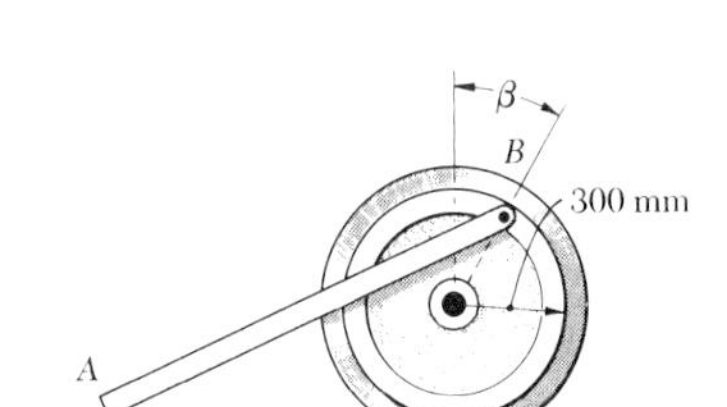

Fig. P15.C2

15.C2 The flanged wheel shown rolls to the right with a constant velocity of 1.5 m/s. Rod AB is 1.2 m long. Express the velocity of A and the angular velocity of the rod as functions of time. Plot the velocity of A and the angular velocity of the rod as functions of time for one revolution of the wheel, setting $t = 0$ when $\beta = 0$.

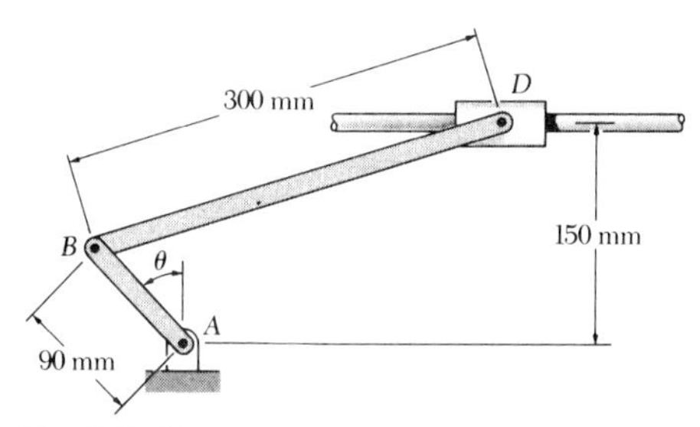

Fig. P15.C3

15.C3 Crank AB has a constant angular velocity of 200 rpm counterclockwise. Express the acceleration of collar D as a function of θ. Plot the acceleration of collar D as a function of time t for one revolution of the crank setting $t = 0$ when $\theta = 0$.

CHAPTER 16
PLANE MOTION OF RIGID BODIES: FORCES AND ACCELERATIONS

SECTION 16.1 to 16.7

16.1 A uniform rod ABC weighs 16 lb and is connected to two collars of negligible weight which slide on horizontal, frictionless rods located in the same vertical plane. If a force **P** of magnitude 4 lb is applied at C, determine (a) the acceleration of the rod, (b) the reactions at B and C.

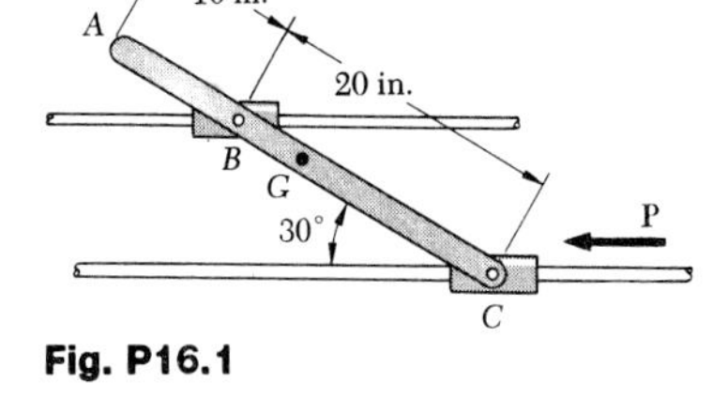

Fig. P16.1

16.2 In Prob. 16.1, determine (a) the required magnitude of **P** if the reaction at B is to be 8 lb upward, (b) the corresponding acceleration of the rod.

16.3 and 16.4 The motion of the 1.5-kg rod AB is guided by two small wheels which roll freely in horizontal slots. If a force **P** of magnitude 5 N is applied at B, determine (a) the acceleration of the rod, (b) the reactions at A and B.

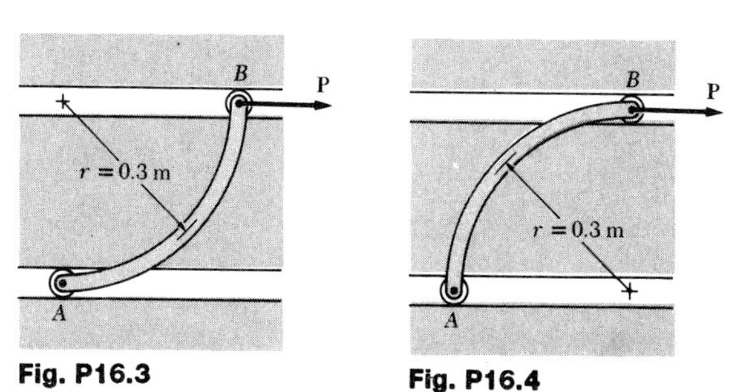

Fig. P16.3 Fig. P16.4

16.5 A uniform rod BC of mass 3 kg is connected to a collar A by a 0.2-m cord AB. Neglecting the mass of the collar and cord, determine (a) the smallest constant acceleration $\mathbf{a}_A$ for which the cord and the rod lie in a straight line, (b) the corresponding tension in the cord.

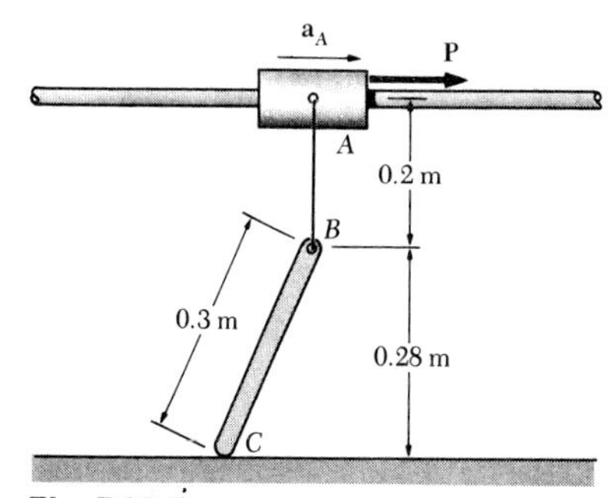

Fig. P16.5

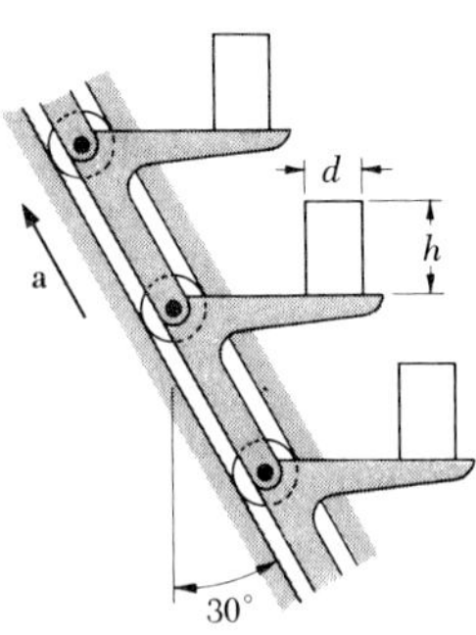

Fig. P16.6

16.6 Cylindrical cans are transported from one elevation to another by the moving horizontal arms shown. Assuming that $\mu_s = 0.25$ between the cans and the arms, determine (*a*) the magnitude of the upward acceleration **a** for which the cans slide on the horizontal arms, (*b*) the smallest ratio h/d for which the cans tip before they slide.

16.7 Solve Prob. 16.6, assuming that the acceleration **a** of the horizontal arms is directed downward.

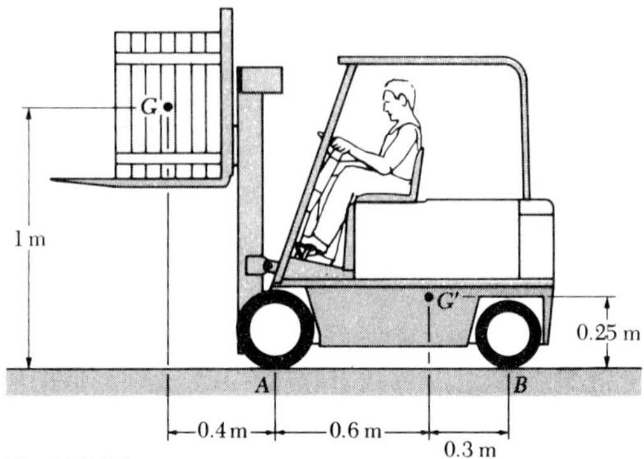

Fig. P16.8

16.8 A 2500-kg fork-lift truck carries the 1200-kg crate at the height shown. The truck is moving to the left when the brakes are applied, causing a deceleration of $3\ \mathrm{m/s^2}$. Knowing that the coefficient of static friction between the crate and the fork lift is 0.60, determine the vertical component of the reaction at each wheel.

16.9 In Prob. 16.8, determine the maximum allowable deceleration of the truck if the crate is not to slide forward and if the truck is not to tip forward.

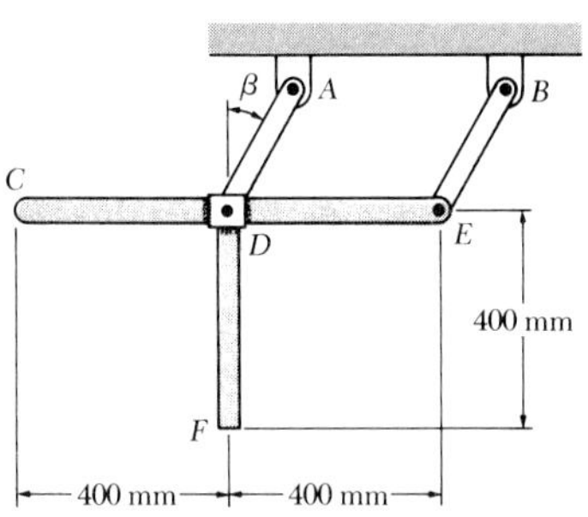

Fig. P16.10 and P16.11

16.10 Three uniform rods *CD*, *DE*, and *DF*, each of mass 1.8 kg, are welded together and are pin-connected to two links *AD* and *BE*. Neglecting the mass of the links, determine the force in each link immediately after the system has been released from rest with $\beta = 30°$.

16.11 Solve Prob. 16.10, knowing that the system has been released from rest with $\beta = 60°$.

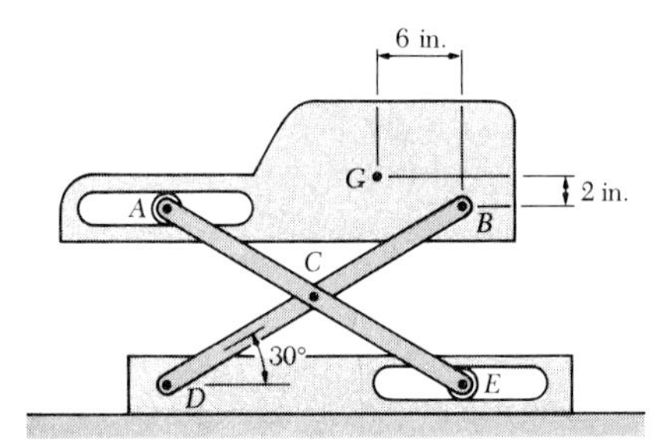

Fig. P16.12

16.12 Members *ACE* and *DCB* are each 24 in. long and are connected by a pin at *C*. The mass center of the 15-lb member *AB* is located at *G*. Determine (*a*) the acceleration of *AB* immediately after the system has been released from rest in the position shown, (*b*) the corresponding force exerted by roller *A* on member *AB*. Neglect the weight of members *ACE* and *DCB*.

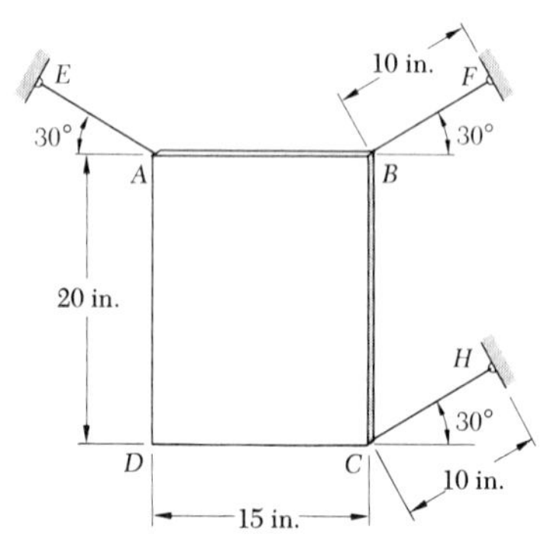

Fig. P16.13

16.13 The thin plate *ABCD* weighs 18 lb and is held in position by the three inextensible wires *AE*, *BF*, and *CH*. Wire *AE* is then cut. Determine (*a*) the acceleration of the plate, (*b*) the tension in wires *BF* and *CH* immediately after wire *AE* has been cut.

16.14 A uniform semicircular plate of mass 5 kg is attached to two links *AB* and *DE*, each 250 mm long, and moves under its own weight. Neglecting the mass of the links and knowing that in the position shown the velocity of the plate is 1.8 m/s, determine the force in each link.

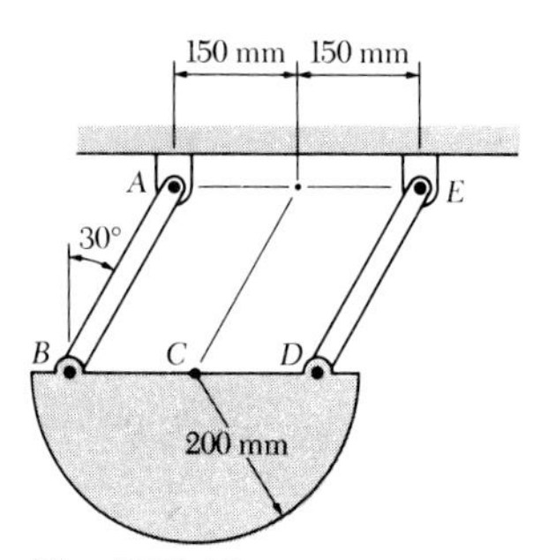

Fig. P16.14

16.15 The T-shaped rod *ABCD* is guided by two pins which slide freely in parallel curved slots of radius 5 in. The rod weighs 4 lb, and its mass center is located at point *G*. Knowing that for the position shown the *horizontal* component of the velocity of *D* is 3 ft/s to the right and the *horizontal* component of the acceleration of *D* is 20 ft/s^2 to the right, determine the magnitude of the force **P**.

16.16 Solve Prob. 16.15, knowing that for the position shown the *horizontal* component of the velocity of *D* is 3 ft/s to the right and the *horizontal* component of the acceleration of *D* is zero.

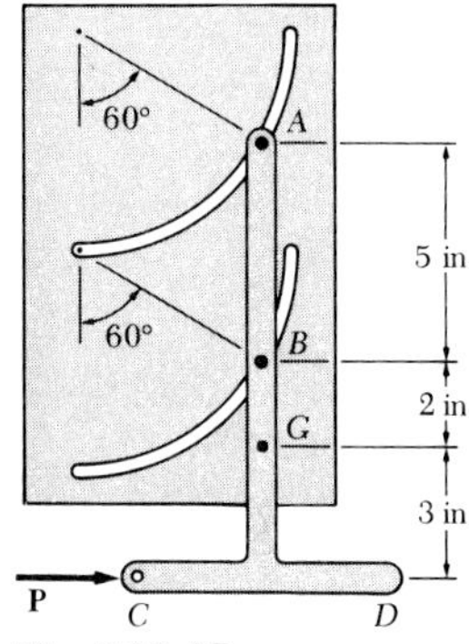

Fig. P16.15

16.17 A 16-lb block is placed on a 4-lb platform *BD* which is held in the position shown by three wires. Determine the accelerations of the block and of the platform immediately after wire *AB* has been cut. Assume that the block (*a*) is rigidly attached to the platform, (*b*) can slide without friction on the platform.

16.18 The coefficients of friction between the 16-lb block and the 4-lb platform *BD* are $\mu_s = 0.50$ and $\mu_k = 0.40$. Determine the accelerations of the block and of the platform immediately after wire *AB* has been cut.

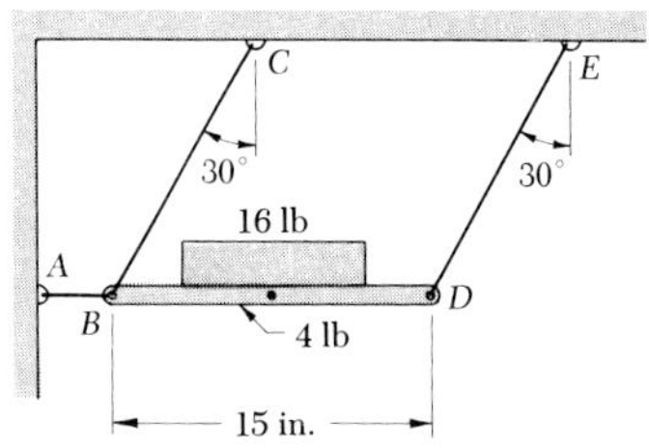

Fig. P16.17 and P16.18

16.19 A turbine-generator unit is shut off when its rotor is rotating at 3600 rpm; it is observed that the rotor coasts to rest in 8.4 min. Knowing that the 1600-kg rotor has a radius of gyration of 220 mm, determine the average magnitude of the couple due to bearing friction.

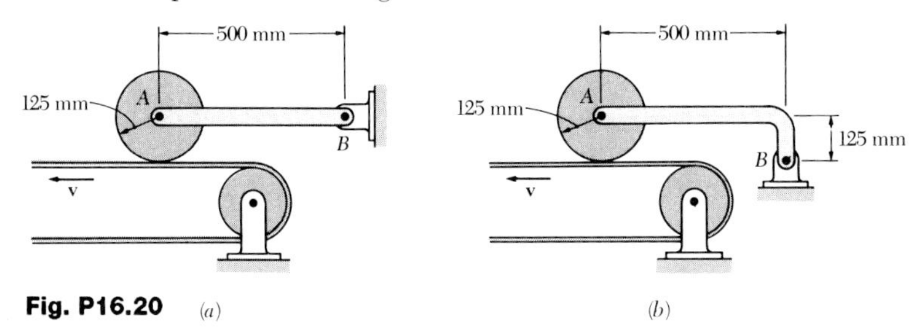

Fig. P16.20 (*a*) (*b*)

16.20 The 4-kg disk is at rest when it is placed in contact with a conveyor belt moving at a constant speed. The link *AB* connecting the center of the disk to the support at *B* is of negligible weight. Knowing that the coefficient of kinetic friction between the disk and the belt is 0.40, determine for each of the arrangements shown the angular acceleration of the disk while slipping occurs.

16.21 Solve Prob. 16.20, assuming that the direction of motion of the conveyor belt is reversed.

16.22 Three disks of the same thickness and same material are attached to a shaft as shown. Disks *A* and *B* are each of mass 3 kg and radius $r = 250$ mm. A couple **M** of magnitude 35 N·m is applied to disk *A* when the system is at rest. Determine the radius *nr* of disk *C* if the angular acceleration of the system is to be 50 rad/s^2.

16.23 Three disks of the same thickness and same material are attached to a shaft as shown. Disks *A* and *B* each have a radius *r*; disk *C* has a radius *nr*. A couple **M** of constant magnitude is applied when the system is at rest. Determine the radius of disk *C* which results in the largest tangential acceleration of a point on the rim of disk *C*.

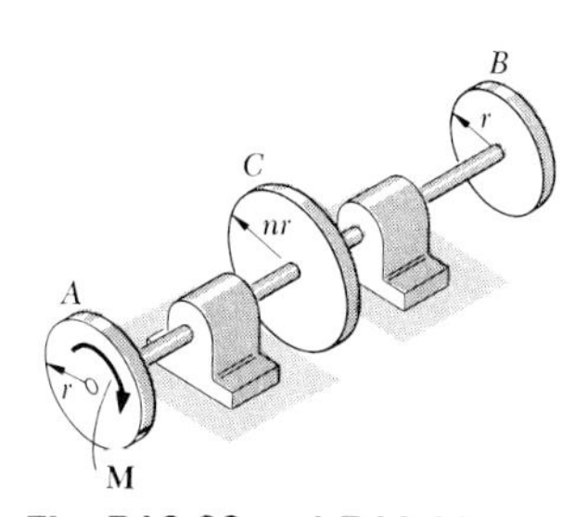

Fig. P16.22 and P16.23

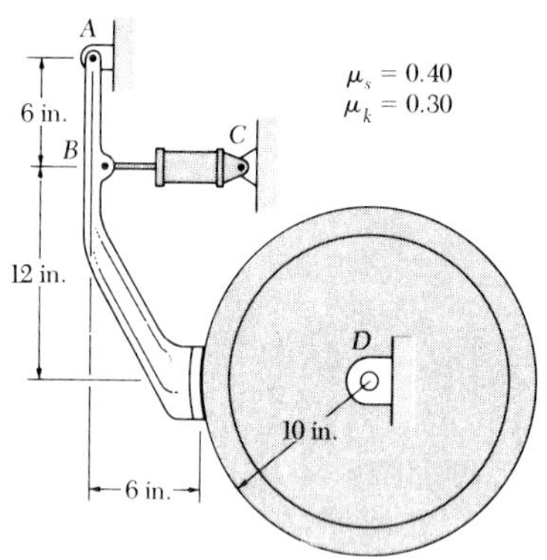

Fig. P16.24

16.24 The 10-in.-radius brake drum is attached to a larger flywheel which is not shown. The total mass moment of inertia of the flywheel and drum is 13.5 lb·ft·s^2. Knowing that the initial angular velocity is 180 rpm clockwise, determine the force which must be exerted by the hydraulic cylinder if the system is to stop in 50 revolutions.

16.25 Solve Prob. 16.24, assuming that the initial angular velocity of the flywheel is 180 rpm counterclockwise.

16.26 The 125-mm-radius brake drum is attached to a flywheel which is not shown. The drum and flywheel together have a mass of 325 kg and a radius of gyration of 725 mm. The coefficient of kinetic friction between the brake band and the drum is 0.40. Knowing that a force **P** of magnitude 50 N is applied at A when the angular velocity is 240 rpm counterclockwise, determine the time required to stop the flywheel when $a = 200$ mm and $b = 250$ mm.

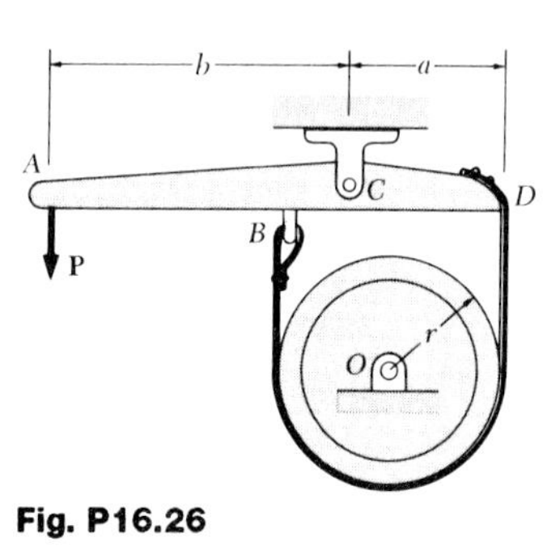

Fig. P16.26

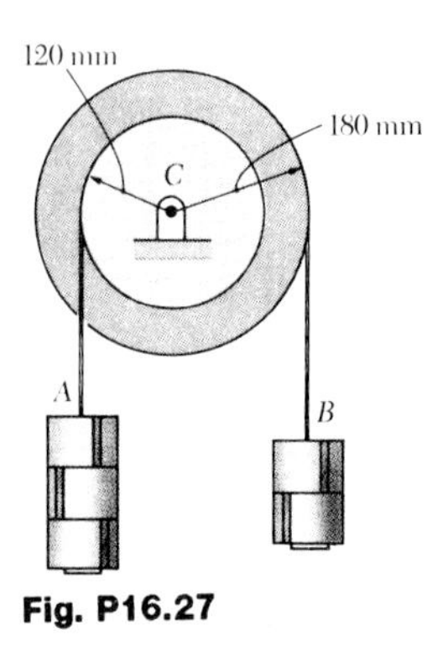

Fig. P16.27

16.27 The double pulley shown has a total mass of 6 kg and a centroidal radius of gyration of 135 mm. Five collars, each of mass 1.2 kg, are attached to cords A and B as shown. When the system is at rest and in equilibrium, one collar is removed from cord A. Neglecting friction, determine (a) the angular acceleration of the pulley, (b) the velocity of cord A at $t = 2.5$ s.

16.28 Solve Prob. 16.27, assuming that one collar is removed from cord B.

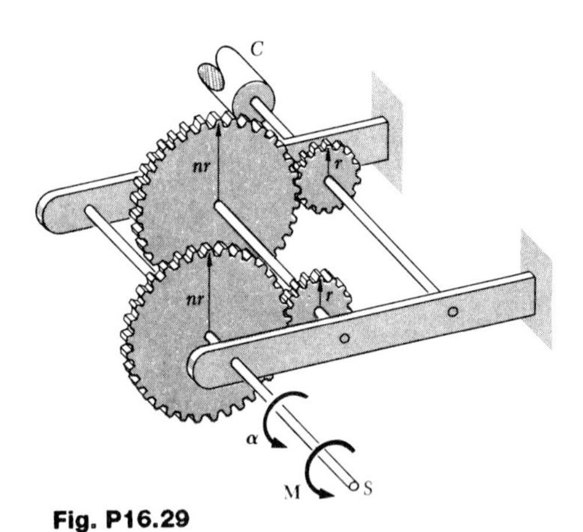

Fig. P16.29

16.29 A coder C, used to record in digital form the rotation of a shaft S, is connected to the shaft by means of the gear train shown, which consists of four gears of the same thickness and of the same material. Two of the gears have a radius r and the other two a radius nr. Let I_R denote the ratio M/α of the moment M of the couple applied to the shaft S and of the resulting angular acceleration α of S. (I_R is sometimes called the "reflected moment of inertia" of the coder and gear train.) Determine I_R in terms of the gear ratio n, the moment of inertia I_0 of the first gear, and the moment of inertia I_C of the coder. Neglect the moments of inertia of the shafts.

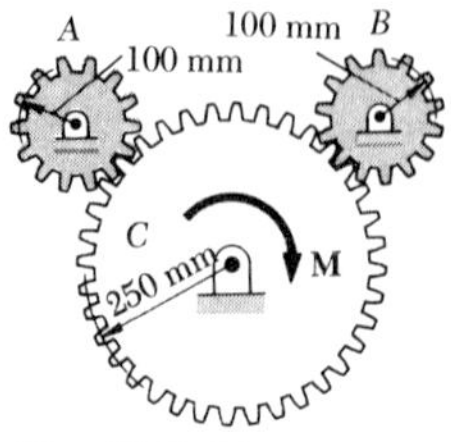

A 100 mm B
100 mm
C
250 mm
M

Fig. P16.30

16.30 Each of the gears A and B has a mass of 2 kg and a radius of gyration of 75 mm; gear C has a mass of 10 kg and a radius of gyration of 225 mm. If a couple **M** of constant magnitude 6 N·m is applied to gear C, determine (a) the angular acceleration of gear A, (b) the tangential force which gear C exerts on gear A.

16.31 Disk A has a mass $m_A = 3$ kg, a radius $r_A = 200$ mm, and an initial angular velocity $\omega_0 = 240$ rpm clockwise. Disk B has a mass $m_B = 1.2$ kg, a radius $r_B = 120$ mm, and is at rest when it is brought into contact with disk A. Knowing that $\mu_k = 0.30$ between the disks and neglecting bearing friction, determine (*a*) the angular acceleration of each disk, (*b*) the reaction at the support C.

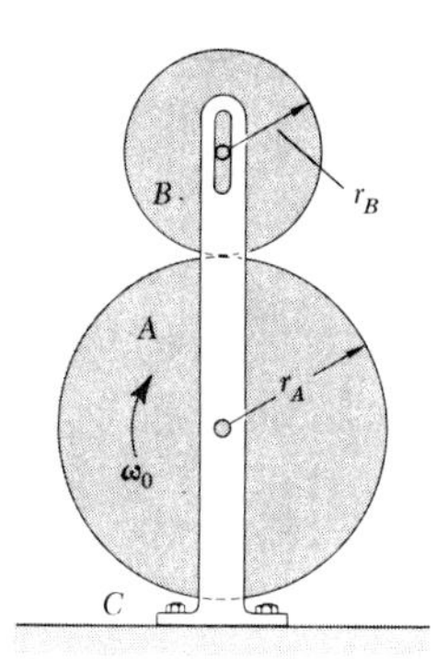

Fig. P16.31 and P16.32

16.32 Disk B is at rest when it is brought into contact with disk A, which has an initial angular velocity ω_0. Show that (*a*) the final angular velocities of the disks are independent of the coefficient of friction μ_k between the disks as long as $\mu_k \neq 0$, (*b*) the final angular velocity of disk A depends only upon ω_0 and the ratio of the masses m_A and m_B of the two disks.

16.33 A cylinder of radius r and mass m rests on two small casters A and B as shown. Initially, the cylinder is at rest and is set in motion by rotating caster B clockwise at high speed so that slipping occurs between the cylinder and caster B. Denoting by μ_k the coefficient of kinetic friction and neglecting the moment of inertia of the free caster A, derive an expression for the angular acceleration of the cylinder.

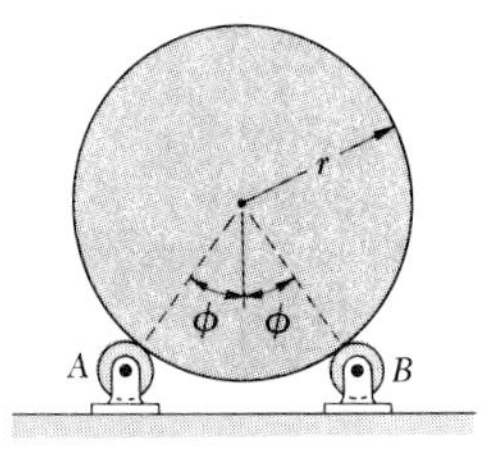

Fig. P16.33

16.34 In Prob. 16.33, assume that no slipping can occur between caster B and the cylinder (such a case would exist if the cylinder and caster had gear teeth along their rims). Derive an expression for the maximum allowable counterclockwise acceleration $\boldsymbol{\alpha}$ of the cylinder if it is not to lose contact with the caster at A.

16.35 The uniform slender rod AB of mass 1.25 kg is at rest on a frictionless horizontal surface. A force $\mathbf{P}$ of magnitude 3 N is applied at A in a horizontal direction perpendicular to the rod. Determine (*a*) the angular acceleration of the rod, (*b*) the acceleration of the center of the rod, (*c*) the point of the rod which has no acceleration.

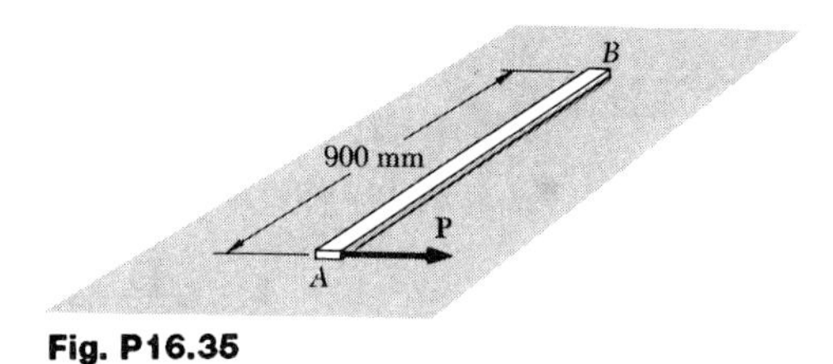

Fig. P16.35

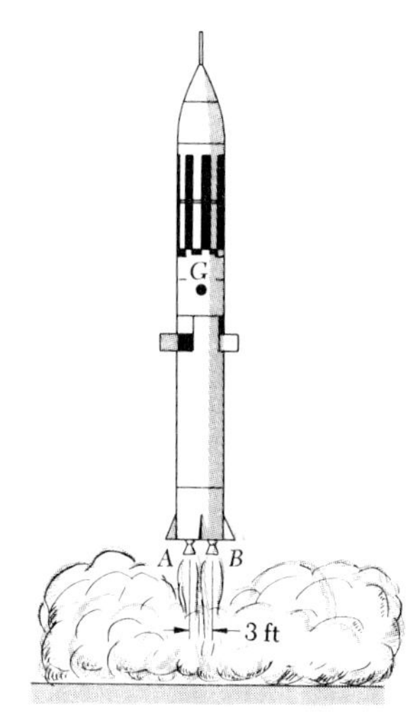

Fig. P16.36

16.36 Shortly after being fired, the experimental rocket shown weighs 25,000 lb and is moving upward with an acceleration of 45 ft/s^2. If at that instant the rocket engine A fails, while rocket engine B continues to operate, determine (*a*) the acceleration of the mass center of the rocket, (*b*) the angular acceleration of the rocket. Assume that the rocket is a uniform slender rod 48 ft long.

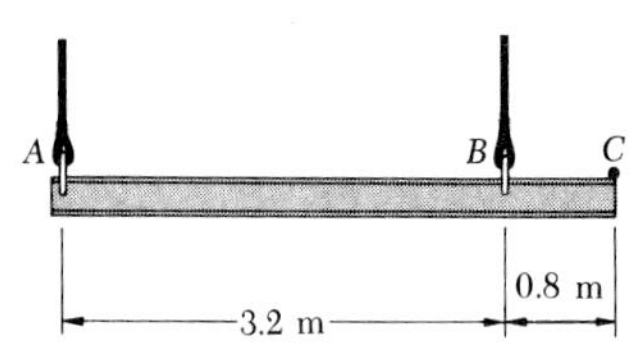

Fig. P16.37

16.37 A 4-m beam of mass 250 kg is lowered from a considerable height by means of two cables unwinding from overhead cranes. As the beam approaches the ground, the crane operators apply brakes to slow the unwinding motion. Determine the acceleration of each cable at that instant, knowing that $T_A = 1000$ N and $T_B = 1800$ N.

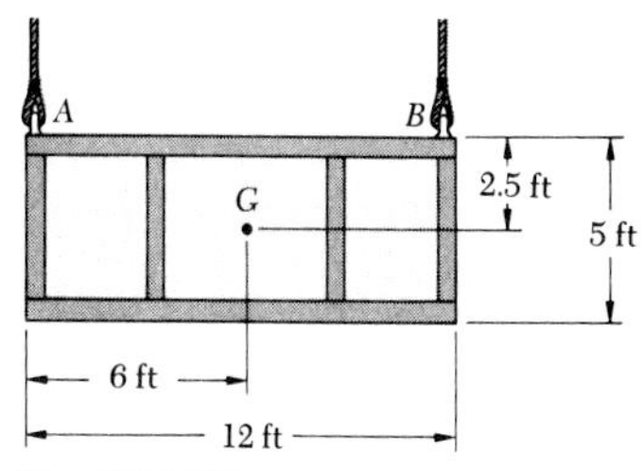

Fig. P16.38

16.38 The 800-lb crate shown is being lowered by means of two overhead cranes. As the crate approaches the ground, the crane operators apply brakes to slow the motion. Determine the acceleration of each cable at that instant, knowing that $T_A = 650$ lb and $T_B = 550$ lb.

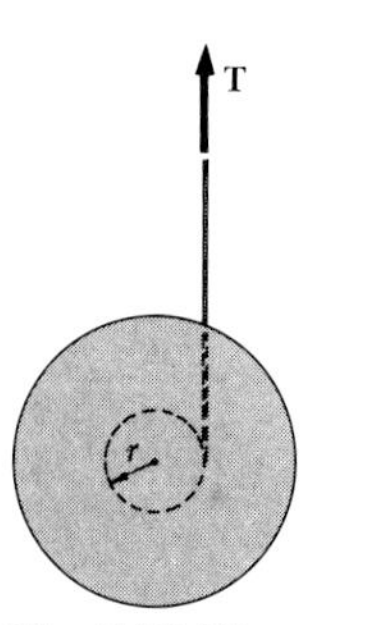

Fig. P16.39

16.39 By pulling on the cord of a yo-yo just fast enough, a person manages to make the yo-yo spin counterclockwise, while remaining at a constant height above the floor. Denoting the weight of the yo-yo by W, the radius of the inner drum on which the cord is wound by r, and the radius of gyration of the yo-yo by $\bar{k}$, determine (a) the tension in the cord, (b) the angular acceleration of the yo-yo.

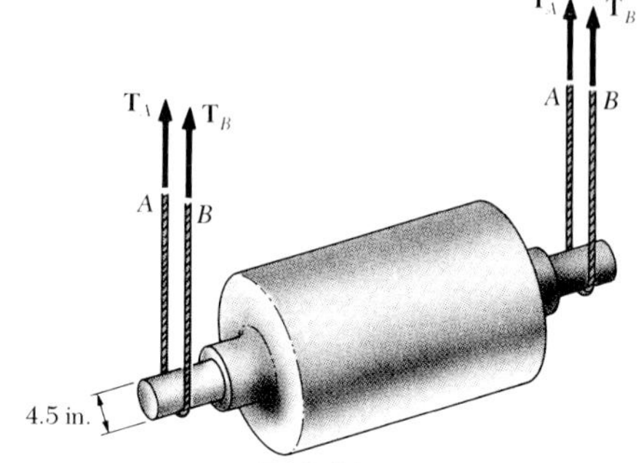

Fig. P16.40 and P16.41

16.40 Structural steel beams are formed by passing them through successive pairs of rolls. The roll shown weighs 2800 lb, has a centroidal radius of gyration of 6 in., and is lifted by two cables looped around its shaft. Knowing that for each cable $T_A = 700$ lb and $T_B = 725$ lb, determine (a) the angular acceleration of the roll, (b) the acceleration of its mass center.

16.41 The steel roll shown weighs 2800 lb, has a centroidal radius of gyration of 6 in., and is being lowered by two cables looped around its shaft. Knowing that at the instant shown the acceleration of the roll is 5 in./s^2 downward and that for each cable $T_A = 680$ lb, determine (a) the corresponding value of the tension T_B, (b) the angular acceleration of the roll.

16.42 and 16.43 A uniform slender bar AB of mass m is suspended from two springs as shown. If spring 2 breaks, determine at that instant (a) the angular acceleration of the bar, (b) the acceleration of point A, (c) the acceleration of point B.

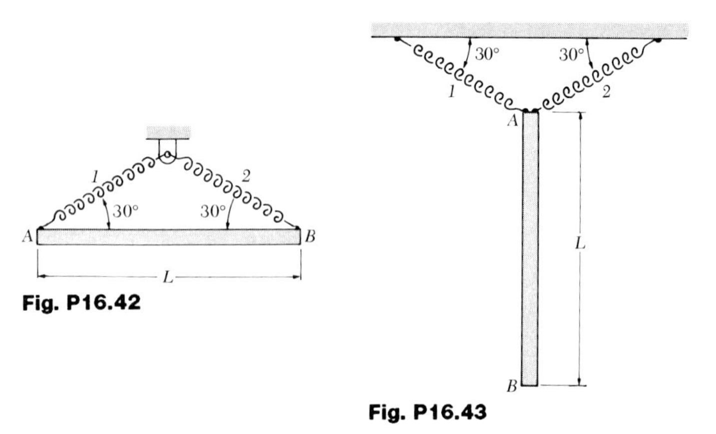

Fig. P16.42

Fig. P16.43

SECTION 16.8

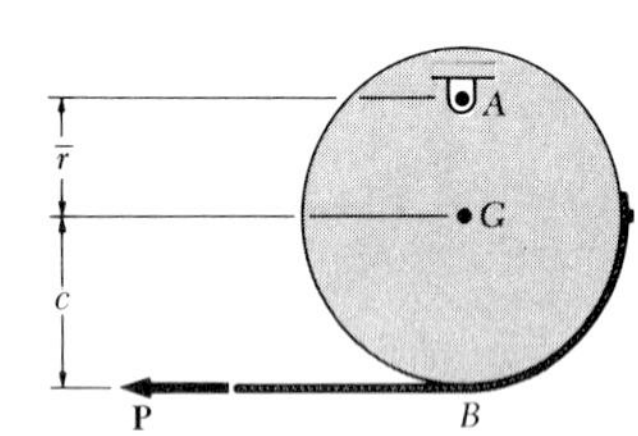

Fig. P16.44

16.44 A uniform disk, of radius $c = 160$ mm and mass $m = 6$ kg, hangs freely from a pin support at A. A force **P** of magnitude 20 N is applied as shown to a cord wrapped around the disk. For $\bar{r} = \frac{3}{4}c = 120$ mm, determine (a) the angular acceleration of the disk, (b) the components of the reaction at A.

16.45 In Prob. 16.44, determine (a) the distance $\bar{r}$ for which the horizontal component of the reaction at A is zero, (b) the corresponding angular acceleration of the disk.

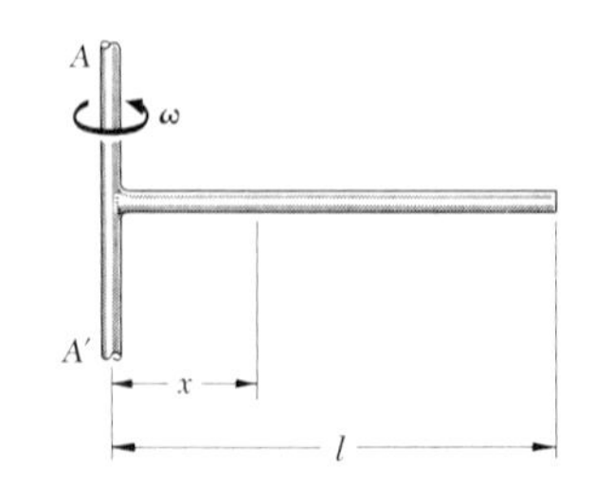

Fig. P16.46

16.46 A uniform slender rod of length l and mass m rotates about a vertical axis AA' at a constant angular velocity ω. Determine the tension in the rod at a distance x from the axis of rotation.

16.47 The rim of a flywheel has a mass of 1800 kg and a mean radius of 600 mm. As the flywheel rotates at a constant angular velocity of 360 rpm, radial forces are exerted on the rim by the spokes and internal forces are developed within the rim. Neglecting the weight of the spokes, determine (*a*) the internal forces in the rim, assuming the radial forces exerted by the spokes to be zero, (*b*) the radial force exerted by each spoke, assuming the tangential forces in the rim to be zero.

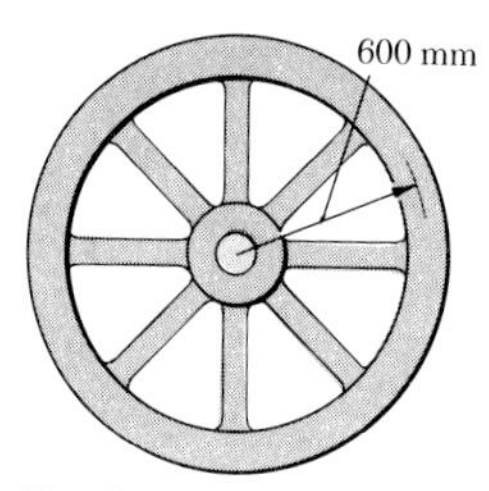

Fig. P16.47

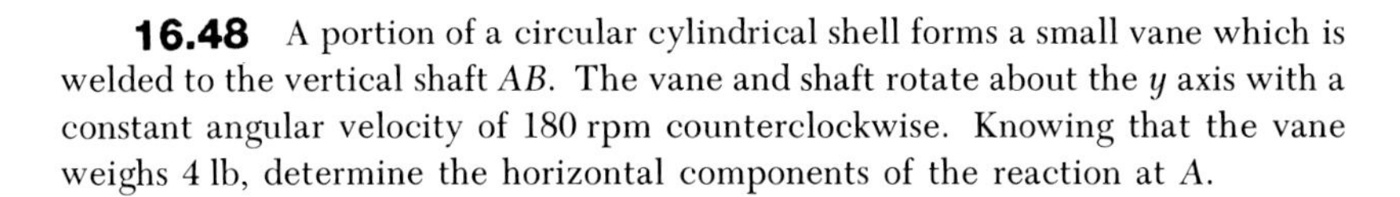

16.48 A portion of a circular cylindrical shell forms a small vane which is welded to the vertical shaft *AB*. The vane and shaft rotate about the *y* axis with a constant angular velocity of 180 rpm counterclockwise. Knowing that the vane weighs 4 lb, determine the horizontal components of the reaction at *A*.

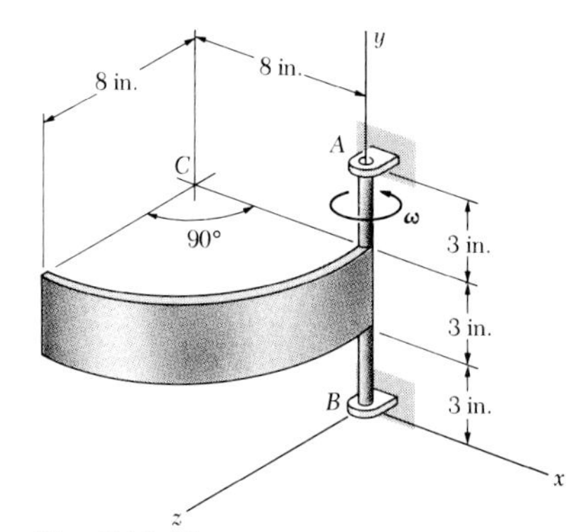

Fig. P16.48

16.49 Centrifugal clutches of the type shown are used to control the operating speed of equipment such as movie cameras and dial telephones. Thin curved members *AB* and *CD* are connected by pins at *A* and *C* to the arm *AC* which may rotate about a fixed point *O*. Each of the members *AB* and *CD* has a mass of 4.5 g and a radius of 10 mm. As the clutch rotates counterclockwise, knobs *H* and *K* slide on the inside of a fixed cylindrical surface of radius 11 mm. Knowing that the coefficient of kinetic friction at *H* and *K* is 0.35 and that the clutch is to have a constant angular velocity of 3000 rpm, determine the couple **M** which must be applied to arm *AC*.

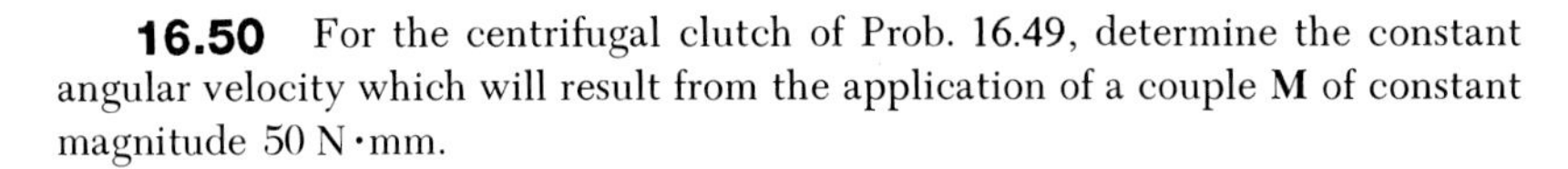

16.50 For the centrifugal clutch of Prob. 16.49, determine the constant angular velocity which will result from the application of a couple **M** of constant magnitude 50 N·mm.

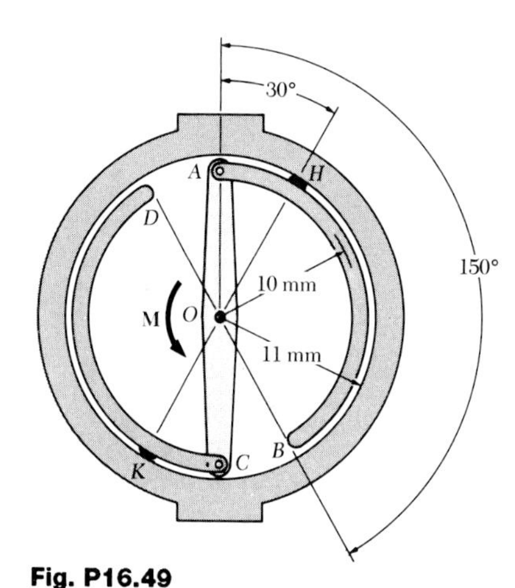

Fig. P16.49

16.51 A uniform square plate of weight *W* is supported as shown. If the cable suddenly breaks, determine (*a*) the angular acceleration of the plate, (*b*) the acceleration of corner *C*, (*c*) the reaction at *A*.

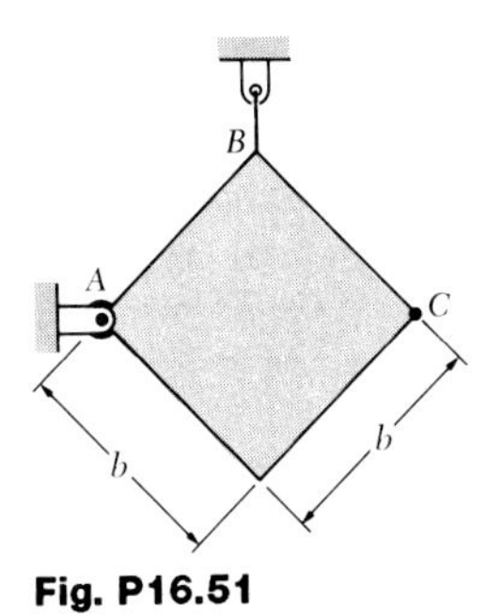

Fig. P16.51

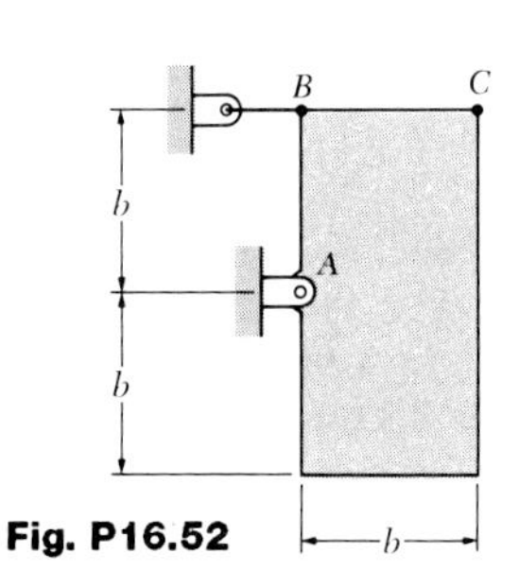

Fig. P16.52

16.52 A uniform rectangular plate of weight *W* is supported as shown. If the cable suddenly breaks, determine (*a*) the angular acceleration of the plate, (*b*) the acceleration of corner *C*, (*c*) the reaction at *A*.

16.53 An 8-lb uniform plate swings freely about *A* in a vertical plane. Knowing that $\mathbf{P} = 0$ and that in the position shown the plate has an angular velocity of 15 rad/s counterclockwise, determine (*a*) the angular acceleration of the plate, (*b*) the components of the reaction at *A*.

16.54 An 8-lb uniform plate rotates about *A* in a vertical plane under the combined effect of gravity and of the vertical force **P**. Knowing that at the instant shown the plate has an angular velocity of 25 rad/s and an angular acceleration of 20 rad/s² both counterclockwise, determine (*a*) the force **P**, (*b*) the components of the reaction at *A*.

Fig. P16.53 and P16.54

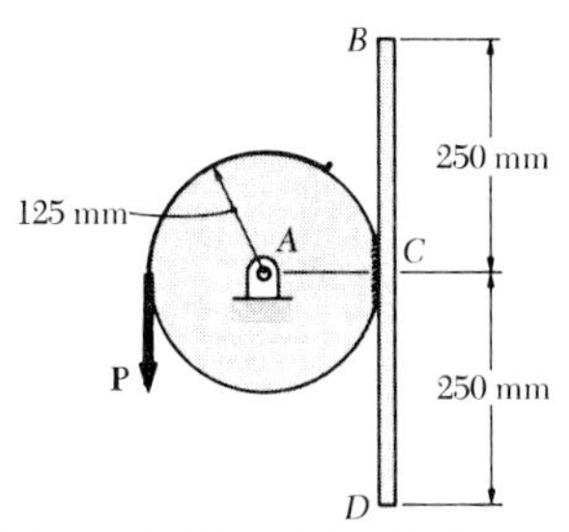

Fig. P16.55 and P16.56

16.55 A 3-kg slender rod is welded to the edge of a 2-kg uniform disk as shown. The assembly rotates about A in a vertical plane under the combined effect of gravity and of the vertical force **P**. Knowing that at the instant shown the assembly has an angular velocity of 12 rad/s and an angular acceleration of 24 rad/s^2 both counterclockwise, determine (a) the force **P**, (b) the components of the reaction at A.

16.56 A 3-kg slender rod is welded to the edge of a 2-kg uniform disk as shown. The assembly swings freely about A in a vertical plane. Knowing that $\mathbf{P} = 0$ and that in the position shown the assembly has an angular velocity of 16 rad/s counterclockwise, determine (a) the angular acceleration of the assembly, (b) the components of the reaction at A.

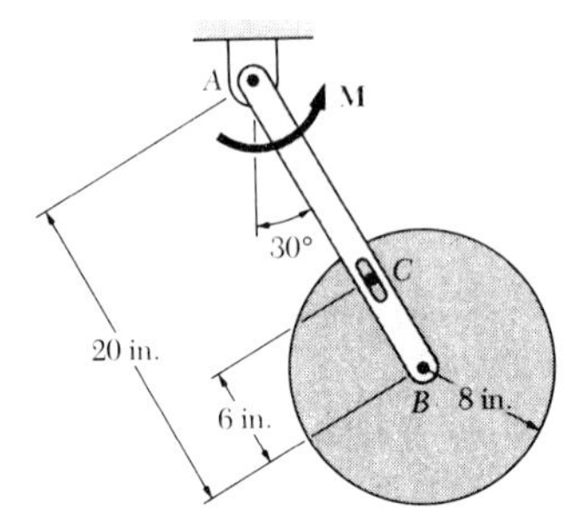

Fig. P16.57

16.57 A 16-lb uniform disk is attached to the 9-lb slender rod AB by means of frictionless pins at B and C. The assembly rotates in a vertical plane under the combined effect of gravity and of a couple **M** which is applied to rod AB. Knowing that at the instant shown the assembly has an angular velocity of 6 rad/s and an angular acceleration of 15 rad/s^2, both counterclockwise, determine (a) the couple **M**, (b) the force exerted by pin C on member AB.

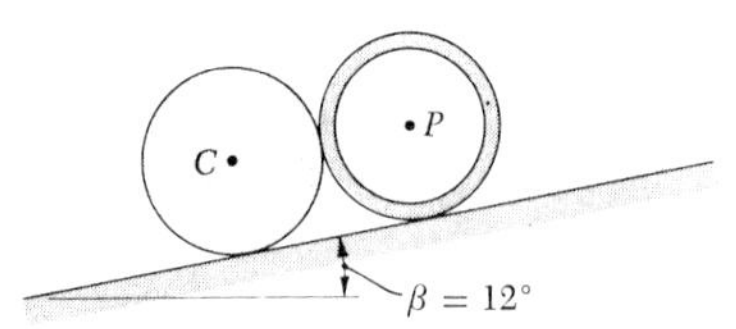

Fig. P16.58

16.58 A homogeneous cylinder C and a section of pipe P are in contact when they are released from rest. Knowing that both the cylinder and the pipe roll without slipping, determine the clear distance between them after 3 s.

16.59 and 16.60 The 9-kg carriage is supported as shown by two uniform disks each of mass 6 kg and radius 150 mm. Knowing that the disks roll without sliding, determine the acceleration of the carriage when a force of 30 N is applied to it.

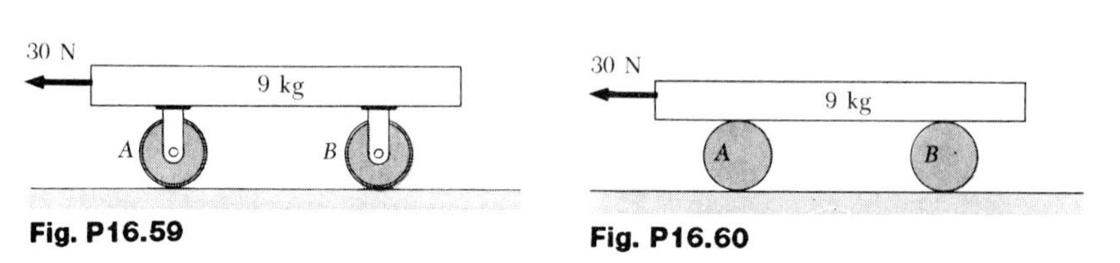

Fig. P16.59

Fig. P16.60

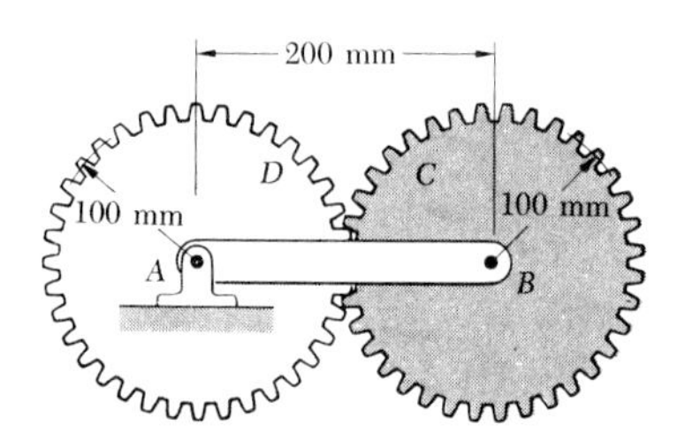

Fig. P16.61

16.61 and 16.62 Gear C has a mass of 3 kg and a centroidal radius of gyration of 75 mm. The uniform bar AB has a mass of 2.5 kg and gear D is stationary. If the system is released from rest in the position shown, determine (a) the angular acceleration of gear C, (b) the acceleration of point B.

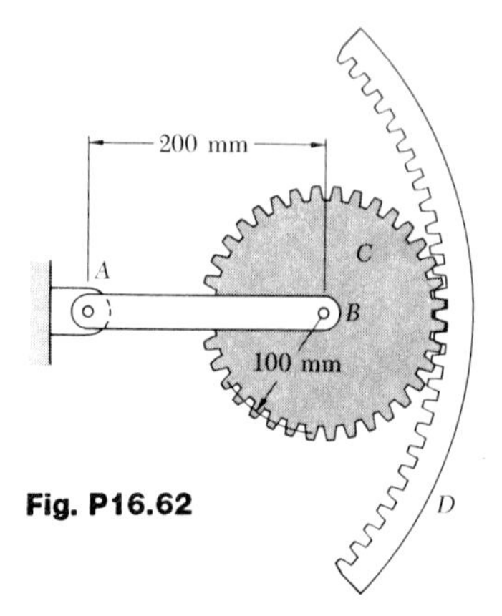

Fig. P16.62

16.63 The disk-and-drum assembly A has a total weight of 15 lb and a centroidal radius of gyration of 6 in. A cord is attached as shown to the assembly A and to the 10-lb uniform disk B. Knowing that the disks roll without sliding, determine the acceleration of the center of each disk for $P = 4$ lb and $Q = 0$.

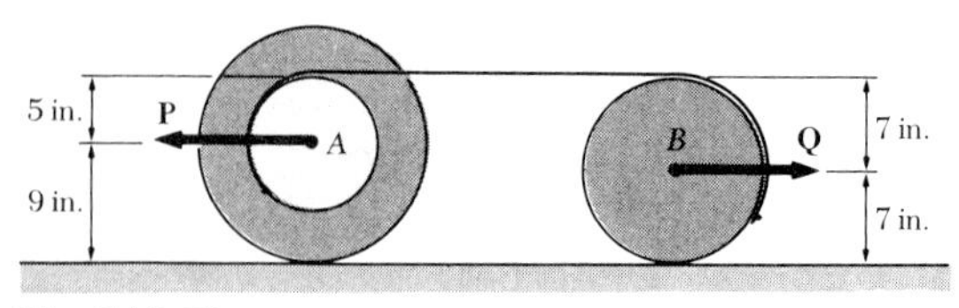

Fig. P16.63

16.64 The mass center G of a 5-kg wheel of radius $R = 300$ mm is located at a distance $r = 100$ mm from its geometric center C. The centroidal radius of gyration is $\bar{k} = 150$ mm. As the wheel rolls without sliding, its angular velocity varies and it is observed that $\omega = 8$ rad/s in the position shown. Determine the corresponding angular acceleration of the wheel.

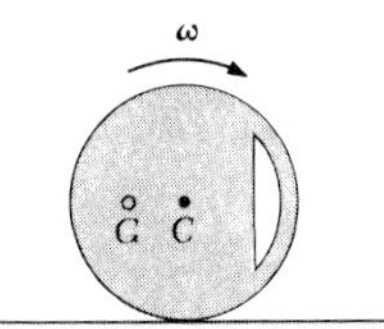

Fig. P16.64

16.65 The motion of the 3-kg uniform rod ACB is guided by two blocks of negligible mass which slide without friction in the slots shown. If the rod is released from rest in the position shown, determine immediately after release (*a*) the angular acceleration of the rod, (*b*) the reaction at A.

16.66 The motion of the 3-kg uniform rod ACB is guided by two blocks of negligible mass which slide without friction in the slots shown. A horizontal force **P** is applied to block A, causing the rod to start from rest with a counterclockwise angular acceleration of 12 rad/s². Determine (*a*) the required force **P**, (*b*) the corresponding reaction at A.

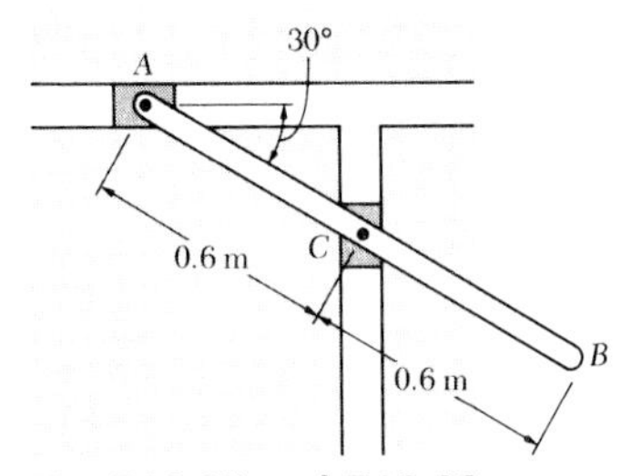

Fig. P16.65 and P16.66

16.67 End B of the 15-lb uniform rod AB rests on a frictionless floor, while end A is attached to a horizontal cable AC. Knowing that at the instant shown the force **P** causes end B of the rod to start from rest with an acceleration of 9 ft/s² to the left, determine (*a*) the force **P**, (*b*) the corresponding tension in cable AC.

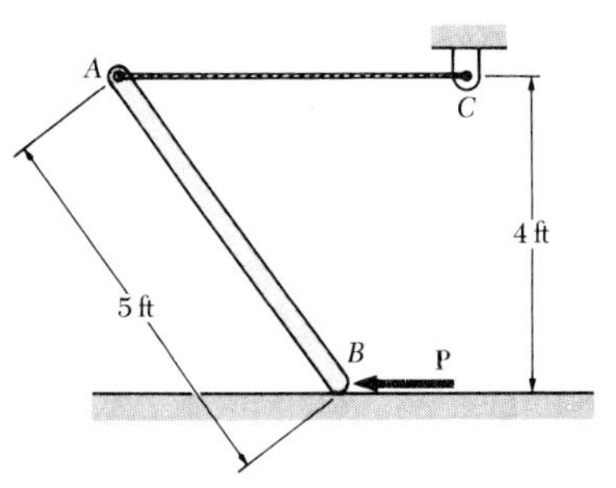

Fig. P16.67

16.68 End A of the 5-kg uniform rod AB rests on the inclined surface, while end B is attached to a collar of negligible mass which may slide along the vertical rod shown. Knowing that the rod is released from rest when $\theta = 35°$ and neglecting the effect of friction, determine immediately after release (*a*) the angular acceleration of the rod, (*b*) the reaction at B.

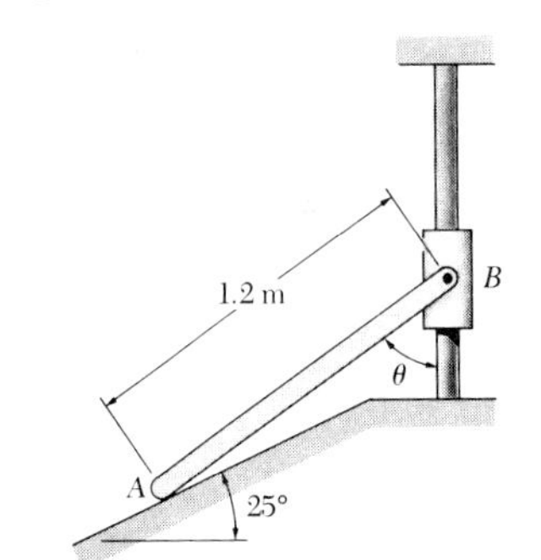

Fig. P16.68

16.69 Rod AB weighs 3 lb and is released from rest in the position shown. Assuming that the ends of the rod slide without friction, determine immediately after release (*a*) the angular acceleration of the rod, (*b*) the reaction at B.

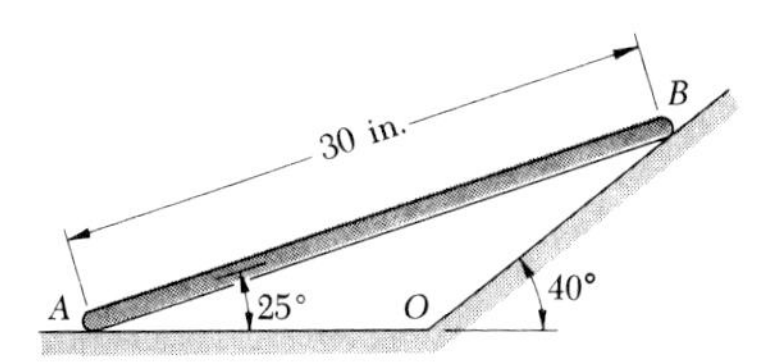

Fig. P16.69

16.70 The motion of a 1.5-kg semicircular rod is guided by two blocks of negligible mass which slide without friction in the slots shown. A horizontal and variable force **P** is applied at B, causing B to move to the right with a constant speed of 5 m/s. For the position shown, determine (*a*) the force **P**, (*b*) the reaction at B.

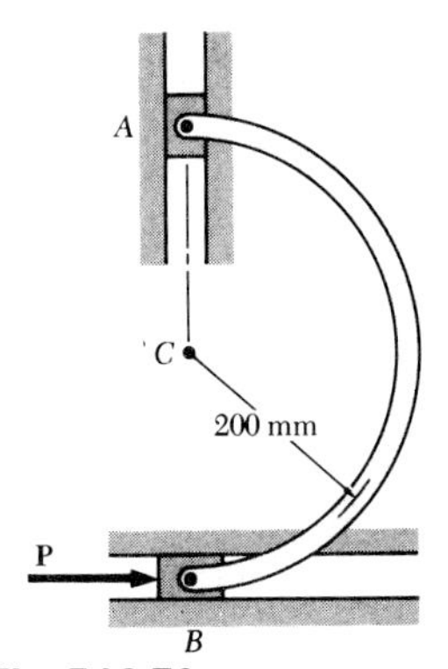

Fig. P16.70

CHAPTER 16 COMPUTER PROBLEMS

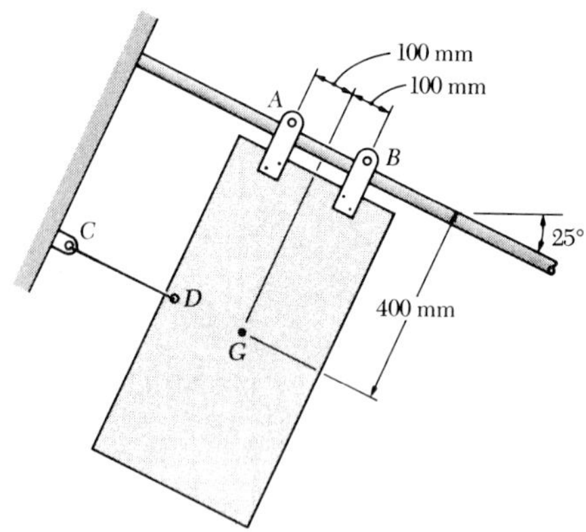

Fig. P16.C1

16.C1 A 20-kg rectangular panel is suspended from two skids A and B and is maintained in the position shown by a wire CD. Derive expressions for the reactions at A and B, and the acceleration of the panel immediately after the wire CD has been cut as functions of the coefficient of kinetic friction μ_k between the skids and the incline track. Plot the reactions at A and B and the acceleration of the panel as a function of μ_k for $\mu_k = 0$ to the maximum value of μ_k for which both skids remain in contact with the track.

16.C2 A completely filled barrel and its contents have a combined mass of 160 kg. A cylinder C of mass m_C is connected to the barrel at a height $h = 700$ mm. The coefficients of friction between the barrel and the floor are $\mu_s = 0.4$ and $\mu_k = 0.35$. Determine and plot the acceleration of the barrel as a function of the mass of cylinder C for the range of values of m_C for which the barrel will not tip.

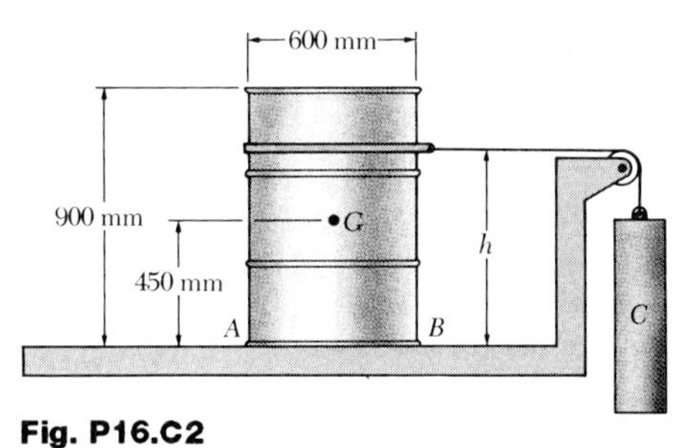

Fig. P16.C2

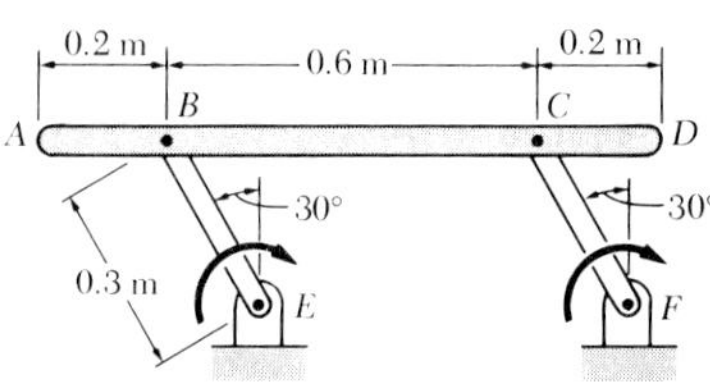

Fig. P16.C3

16.C3 The cranks BE and CF are made to rotate at constant speed of 90 rpm. Determine and plot the vertical components of the forces exerted on the 6-kg uniform rod $ABCD$ by the pins B and C for one revolution of the cranks.

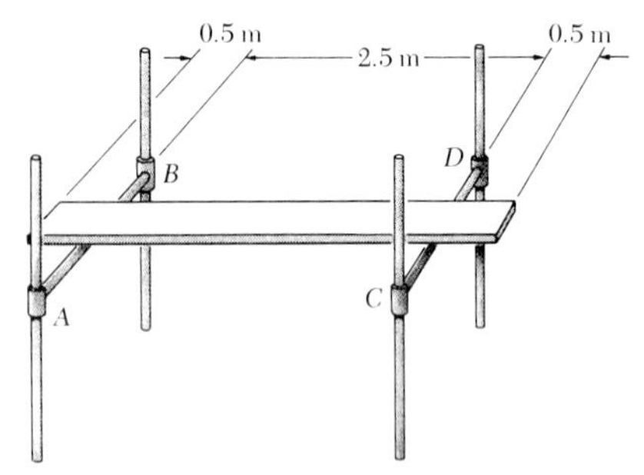

Fig. P16.C4

16.C4 A 3.5-m plank of mass 30 kg rests on two horizontal pipes AB and CD of a scaffolding. The pipes are 2.5 m apart, and the plank overhangs 0.5 m at each end. A 70-kg worker is standing on the plank when pipe CD suddenly breaks. Determine and plot the initial acceleration of the worker as a function of his position along the plank.

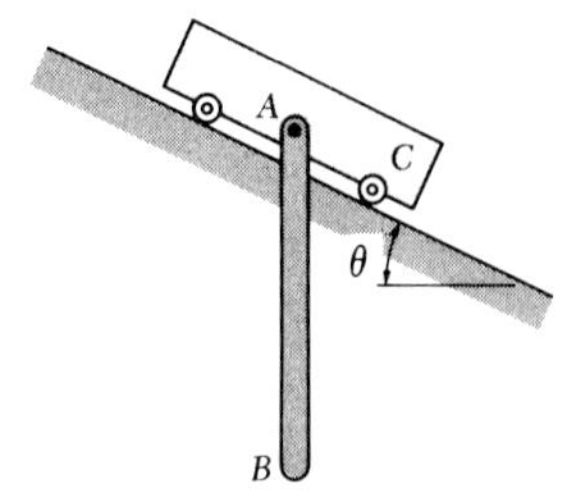

Fig. P16.C5

16.C5 The uniform rod AB, of weight 20 lb and length 3 ft, is attached to the 30-lb cart C. Determine and plot the acceleration of the cart and the angular acceleration of the rod as functions of the angle θ, for $\theta = 0$ to $\theta = 90^\circ$, immediately after the system has been released from rest. Neglect friction.

CHAPTER 17
PLANE MOTION OF RIGID BODIES: ENERGY AND MOMENTUM METHODS

SECTIONS 17.1 to 17.7

17.1 Three disks of the same thickness and same material are attached to a shaft as shown. Disks A and B each have a radius r; disk C has a radius nr. A couple $\mathbf{M}$ of constant magnitude is applied when the system is at rest and is removed after the system has executed one revolution. Determine the radius of disk C which results in the largest final speed of a point on the rim of disk C.

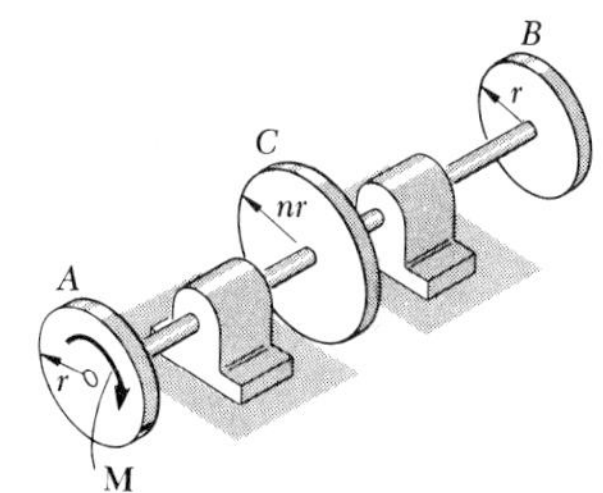

Fig. P17.1 and P17.2

17.2 Three disks of the same thickness and same material are attached to a shaft as shown. Disks A and B each have a mass of 8 kg and a radius $r = 240$ mm. A couple $\mathbf{M}$ of magnitude 40 N·m is applied to disk A when the system is at rest. Determine the radius nr of disk C if the angular velocity of the system is to be 900 rpm after 25 revolutions.

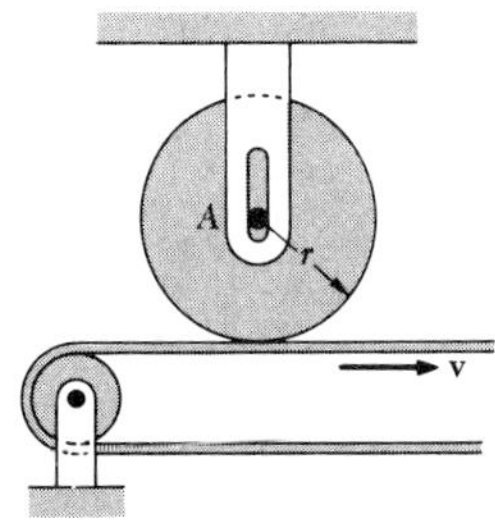

Fig. P17.3

17.3 Disk A has a mass of 4 kg and a radius $r = 90$ mm; it is at rest when it is placed in contact with the belt, which moves with a constant speed $v = 15$ m/s. Knowing that $\mu_k = 0.25$ between the disk and the belt, determine the number of revolutions executed by the disk before it reaches a constant angular velocity.

17.4 Gear G weighs 8 lb and has a radius of gyration of 2.4 in., while gear D weighs 30 lb and has a radius of gyration of 4.8 in. A couple $\mathbf{M} = (20 \text{ lb}\cdot\text{in.})\mathbf{i}$ is applied to shaft CE. Neglecting the mass of shafts CE and FH, determine (a) the number of revolutions of gear G required for its angular velocity to increase from 240 to 720 rpm, (b) the corresponding tangential force acting on gear G.

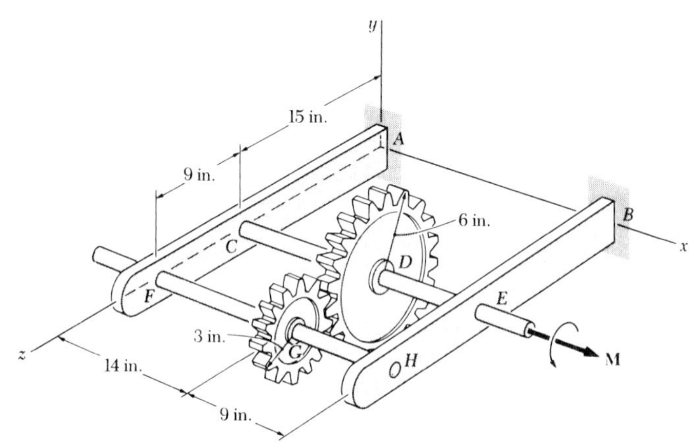

Fig. P17.4

17.5 Solve Prob. 17.4, assuming that the couple $\mathbf{M} = (20 \text{ lb}\cdot\text{in.})\mathbf{i}$ is applied to shaft FH.

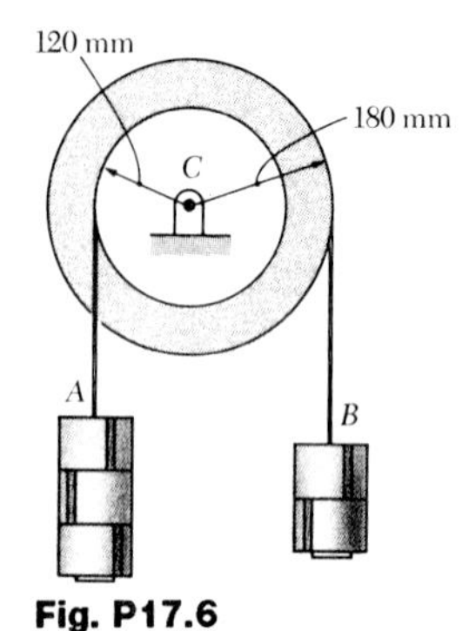

Fig. P17.6

17.6 The double pulley shown has a total mass of 6 kg and a centroidal radius of gyration of 135 mm. Five collars, each of mass 1.2 kg, are attached to cords A and B as shown. When the system is at rest and in equilibrium, one collar is removed from cord A. Knowing that the bearing friction is equivalent to a couple $\mathbf{M}$ of magnitude 0.5 N·m, determine the velocity of cord A after it has moved 600 mm.

17.7 Solve Prob. 17.6, assuming that one collar is removed from cord B.

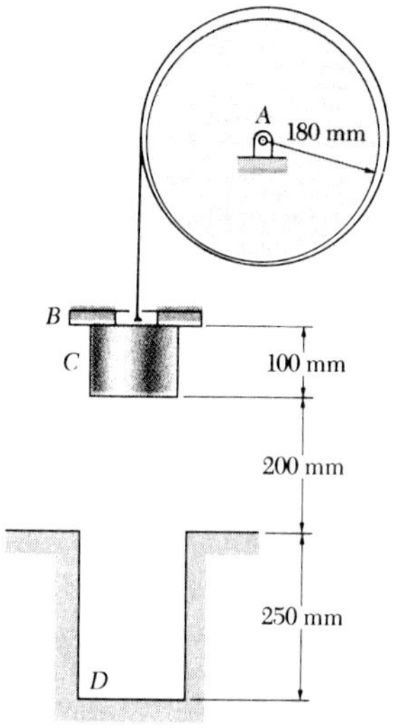

Fig. P17.8

17.8 The pulley shown has a mass of 5 kg and a radius of gyration of 150 mm. The 3-kg cylinder C is attached to a cord which is wrapped around the pulley as shown. A 1.8-kg collar B is then placed on the cylinder and the system is released from rest. After the cylinder has moved 300 mm, the collar is removed and the cylinder continues to move downward into a pit. Determine the velocity of cylinder C just before it strikes the bottom D of the pit.

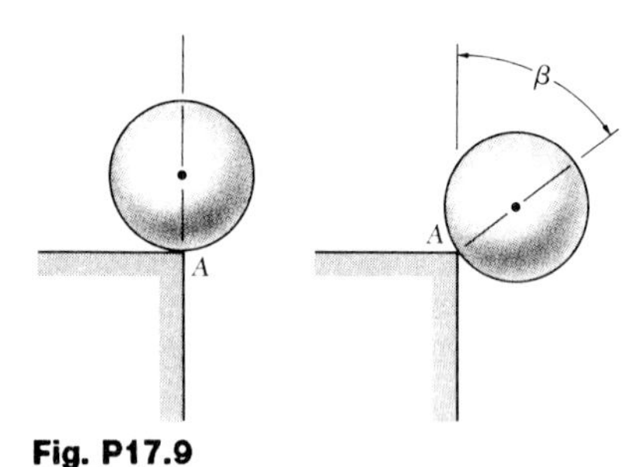

Fig. P17.9

17.9 A uniform sphere of radius r is placed at corner A and is given a slight clockwise motion. Assuming that the corner is sharp and becomes slightly embedded in the sphere, so that the coefficient of static friction at A is very large, determine (a) the angle β through which the sphere will have rotated when it loses contact with the corner, (b) the corresponding velocity of the center of the sphere.

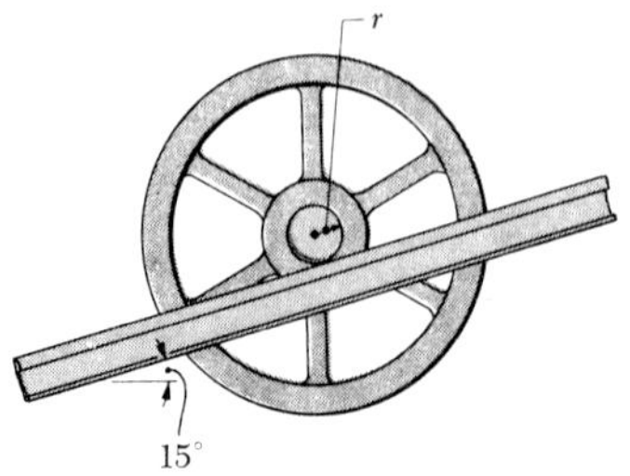

Fig. P17.10 and P17.11

17.10 A flywheel of centroidal radius of gyration $\bar{k} = 25$ in. is rigidly attached to a shaft of radius $r = 1.25$ in. which may roll along parallel rails. Knowing that the system is released from rest, determine the velocity of the center of the shaft after it has moved 10 ft.

17.11 A flywheel is rigidly attached to a 40-mm-radius shaft which rolls without sliding along parallel rails. The system is released from rest and attains a speed of 160 mm/s after moving 1.5 m along the rails. Determine the centroidal radius of gyration of the system.

17.12 The mass center G of a 5-kg wheel of radius $R = 300$ mm is located at a distance $r = 100$ mm from its geometric center C. The centroidal radius of gyration of the wheel is $\bar{k} = 150$ mm. As the wheel rolls without sliding, its angular velocity is observed to vary. Knowing that $\omega = 6$ rad/s in the position shown, determine (*a*) the angular velocity of the wheel when the mass center G is directly above the geometric center C, (*b*) the reaction at the horizontal surface at the same instant.

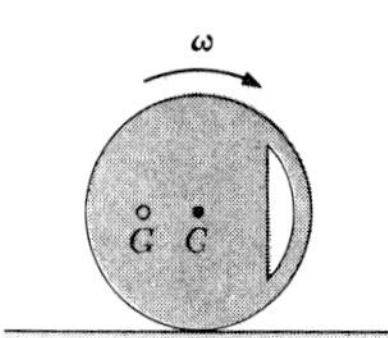

Fig. P17.12

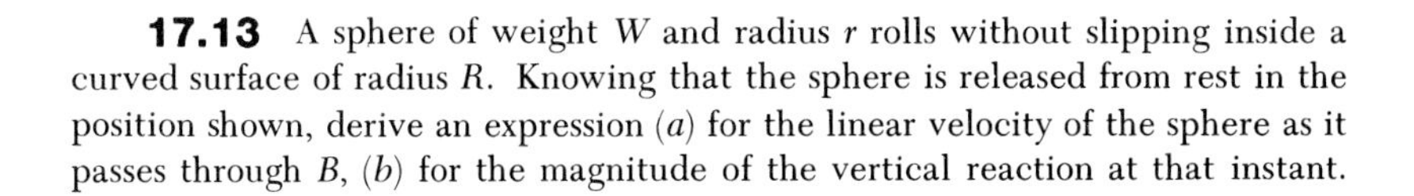

17.13 A sphere of weight W and radius r rolls without slipping inside a curved surface of radius R. Knowing that the sphere is released from rest in the position shown, derive an expression (*a*) for the linear velocity of the sphere as it passes through B, (*b*) for the magnitude of the vertical reaction at that instant.

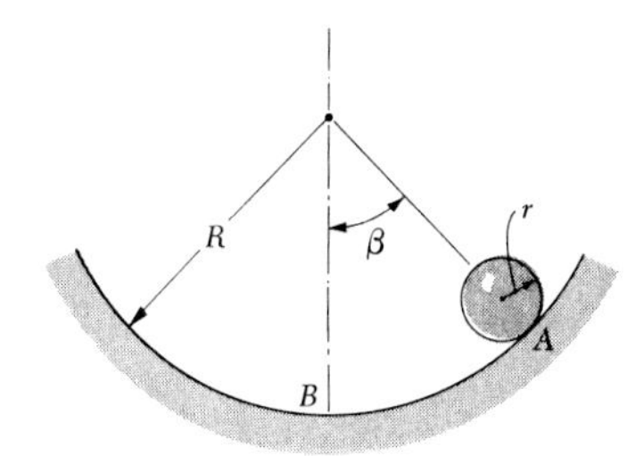

Fig. P17.13

17.14 Two uniform cylinders, each of weight $W = 14$ lb and radius $r = 5$ in., are connected by a belt as shown. Knowing that at the instant shown the angular velocity of cylinder B is 30 rad/s clockwise, determine (*a*) the distance through which cylinder A will rise before the angular velocity of cylinder B is reduced to 5 rad/s, (*b*) the tension in the portion of belt connecting the two cylinders.

17.15 Two uniform cylinders, each of weight $W = 14$ lb and radius $r = 5$ in., are connected by a belt as shown. If the system is released from rest, determine (*a*) the velocity of the center of cylinder A after it has moved through 3 ft, (*b*) the tension in the portion of belt connecting the two cylinders.

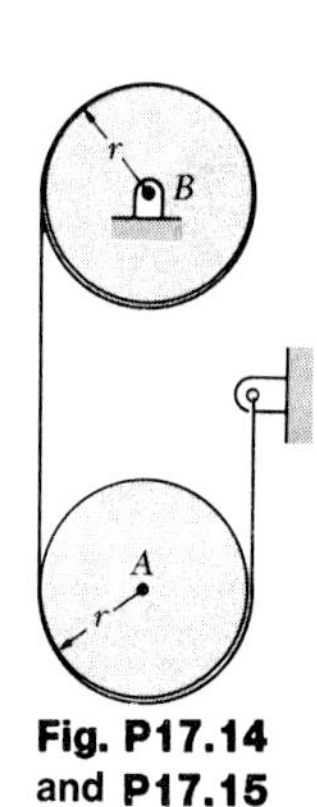

Fig. P17.14 and P17.15

17.16 The 8-kg rod AB is attached by pins to two 5-kg uniform disks as shown. The assembly rolls without sliding on a horizontal surface. If the assembly is released from rest when $\theta = 60°$, determine (*a*) the angular velocity of the disks when $\theta = 180°$, (*b*) the force exerted by the surface on each disk at that instant.

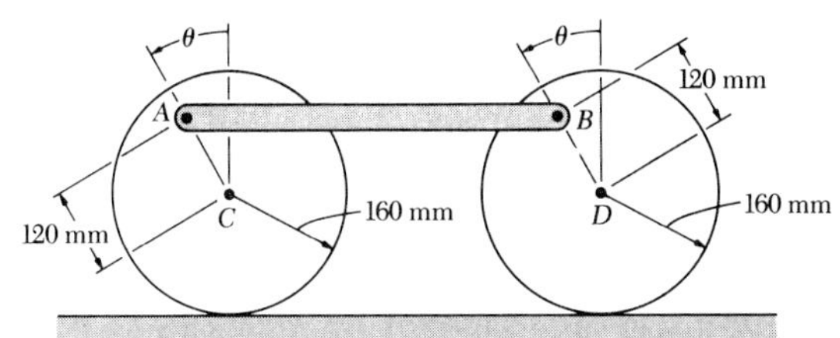

Fig. P17.16 and P17.17

17.17 The 8-kg rod AB is attached by pins to two 5-kg uniform disks as shown. The assembly rolls without sliding on a horizontal surface. Knowing that the velocity of rod AB is 700 mm/s to the left when $\theta = 0$, determine (*a*) the velocity of rod AB when $\theta = 180°$, (*b*) the force exerted by the surface on each disk at that instant.

17.18 The uniform rods AB and BC are of mass 2 kg and 5 kg, respectively, and collar C has a mass of 3 kg. If the system is released from rest in the position shown, determine the velocity of point B after rod AB has rotated through 90°.

17.19 The uniform rods AB and BC are of mass 2 kg and 5 kg, respectively, and collar C has a mass of 3 kg. Knowing that at the instant shown the velocity of collar C is 0.8 m/s downward, determine the velocity of point B after rod AB has rotated through 90°.

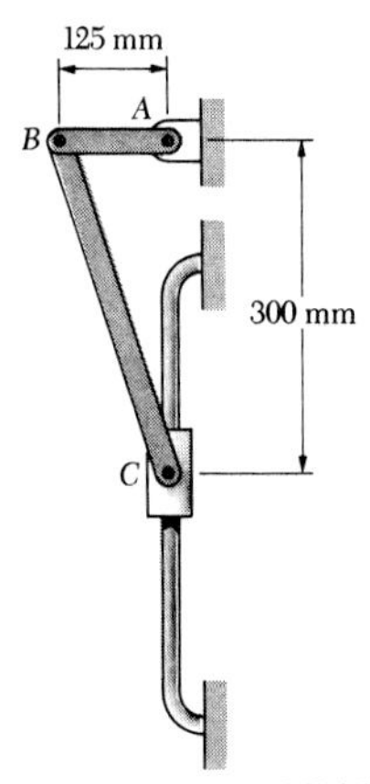

Fig. P17.18 and P17.19

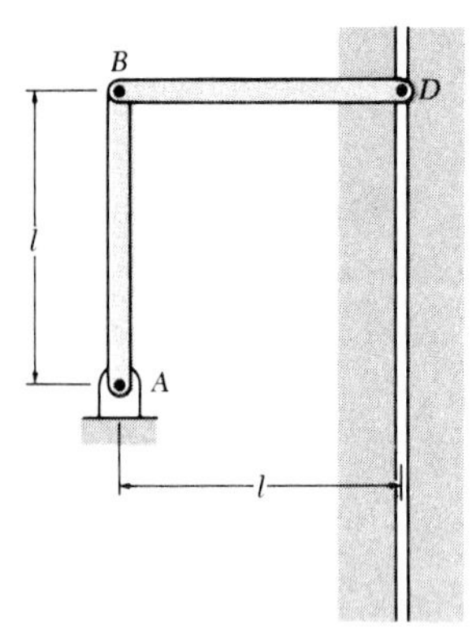

Fig. P17.20

17.20 Two uniform rods, each of mass m and length l, are connected to form the linkage shown. End D of rod BD may slide freely in the vertical slot, while end A of rod AB is attached to a fixed pin support. If the system is released from rest in the position shown, determine the velocity of end D at the instant when (*a*) ends A and D are at the same elevation, (*b*) rod AB is horizontal.

17.21 Wheel A weighs 9 lb, has a centroidal radius of gyration of 6 in., and rolls without sliding on the horizontal surface. Each of the uniform rods AB and BC is 20 in. long and weighs 5 lb. If point A is moved slightly to the left and released, determine the velocity of point A as rod BC passes through a horizontal position.

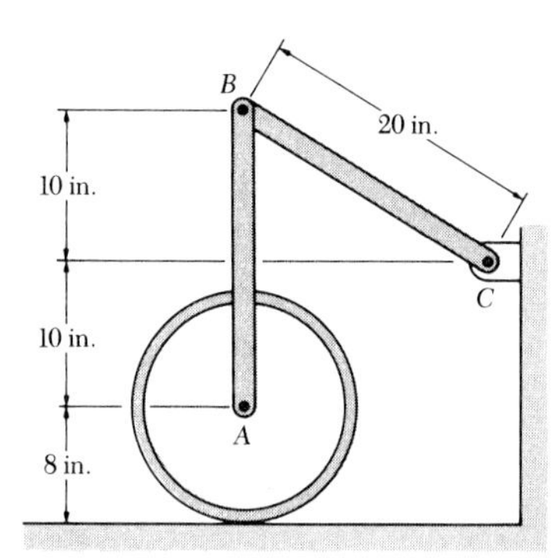

Fig. P17.21

17.22 The motor shown runs a machine attached to the shaft at A. The motor develops 5 hp and runs at a constant speed of 360 rpm. Determine the magnitude of the couple exerted (*a*) by the shaft on pulley A, (*b*) by the motor on pulley B.

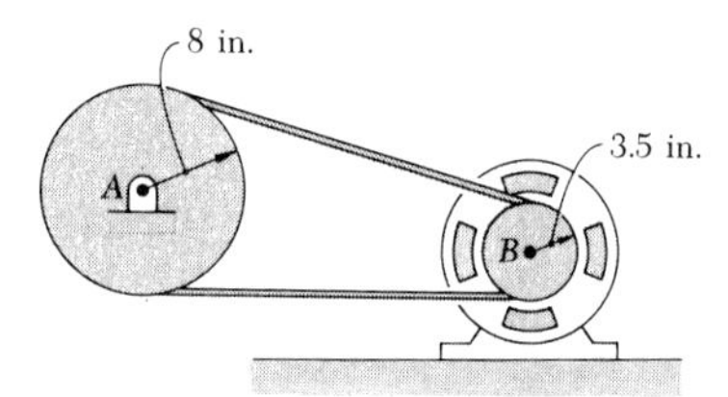

Fig. P17.22

17.23 Three shafts and four gears are used to form a gear train which will transmit 7.5 kW from the motor at A to a machine tool at F. (Bearings for the shafts are omitted in the sketch.) Knowing that the frequency of the motor is 30 Hz, determine the magnitude of the couple which is applied to shaft (*a*) AB, (*b*) CD, (*c*) EF.

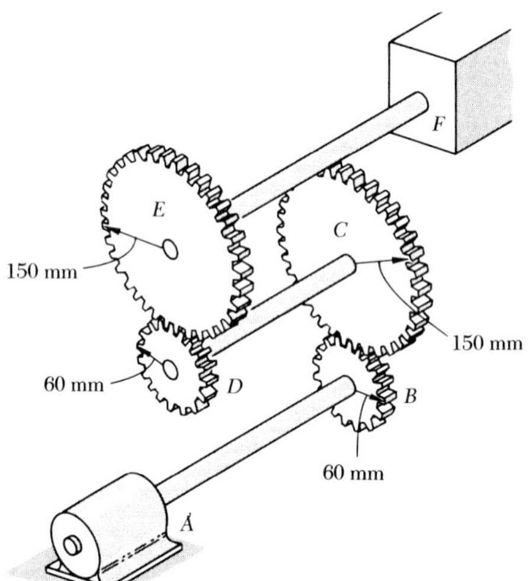

Fig. P17.23

SECTION 17.8 to 17.10

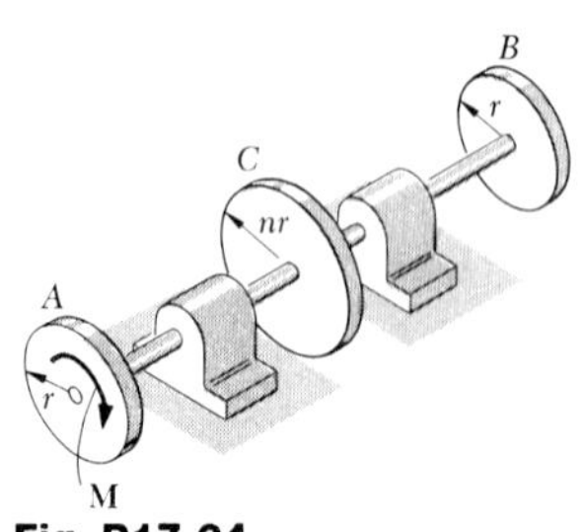

Fig. P17.24

17.24 Three disks of the same thickness and same material are attached to a shaft as shown. Disks A and B each have a mass of 8 kg and a radius $r =$ 240 mm. A couple $\mathbf{M}$ of magnitude 40 N·m is applied to disk A when the system is at rest. Determine the radius nr of disk C if the angular velocity of the system is to be 900 rpm after 3 s.

17.25 A bolt located 2 in. from the center of an automobile wheel is tightened by applying the couple shown for 0.10 s. Assuming that the wheel is free to rotate and is initially at rest, determine the resulting angular velocity of the wheel. The wheel weighs 42 lb and has a radius of gyration of 11 in.

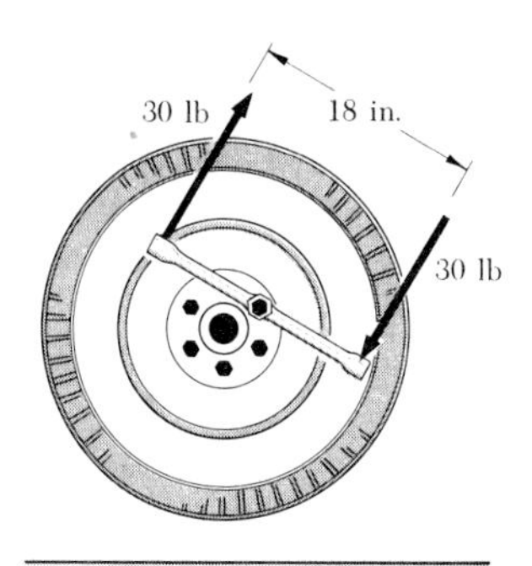

Fig. P17.25

17.26 Each of the double pulleys shown has a centroidal mass moment of inertia of 0.10 kg·m², an inner radius of 80 mm, and an outer radius of 120 mm. Neglecting bearing friction, determine (*a*) the velocity of the cylinder 3 s after the system is released from rest, (*b*) the tension in the cord connecting the pulleys.

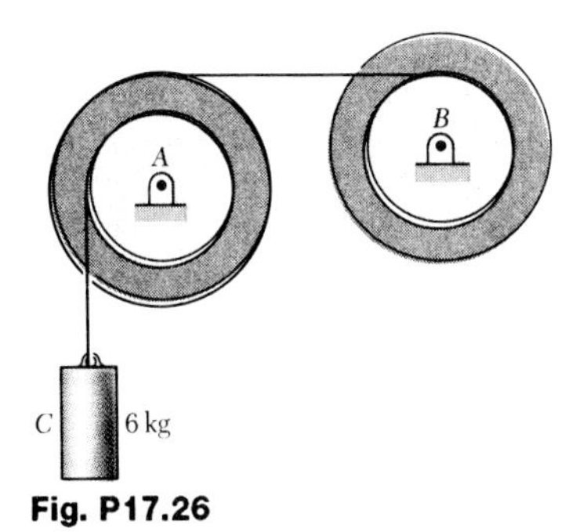

Fig. P17.26

17.27 Solve Prob. 17.26, assuming that the bearing friction at A and at B is equivalent to a couple of magnitude 0.25 N·m.

17.28 A drum of 60-mm radius is attached to a disk of 120-mm radius. The disk and the drum have a total mass of 4 kg and a radius of gyration of 90 mm. A cord is wrapped around the drum and pulled with a force **P** of magnitude 15 N. Knowing that the disk is initially at rest, determine (*a*) the velocity of the center G after 2.5 s, (*b*) the friction force required to prevent slipping.

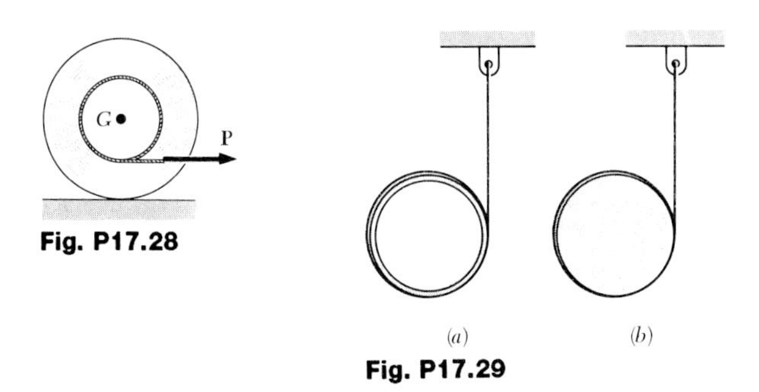

Fig. P17.28

Fig. P17.29

17.29 Cords are wrapped around a thin-walled pipe and a solid cylinder as shown. Knowing that the pipe and the cylinder are each released from rest at time $t = 0$, determine at time t the velocity of the center of (*a*) the pipe, (*b*) the cylinder.

17.30 The double pulley shown weighs 6 lb and has a centroidal radius of gyration of 3 in. The 2-in.-radius inner pulley is rigidly attached to the 4-in.-radius outer pulley, and the motion of the center A is guided by a smooth pin at A which slides in a vertical slot. When the pulley is at rest, a force **P** of magnitude 8 lb is applied to cord C. Determine the velocity of the center of the pulley after 5 s.

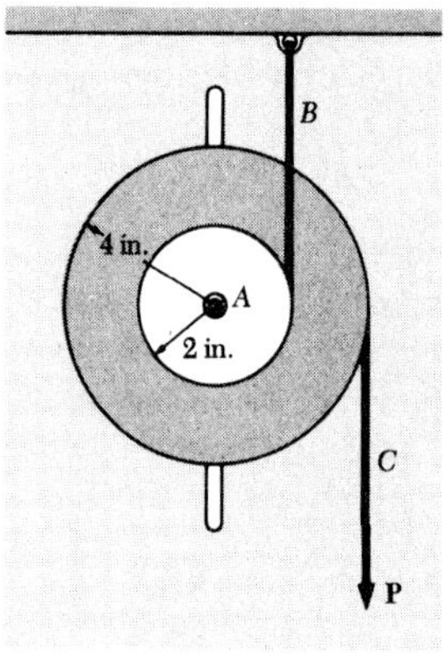

Fig. P17.30

17.31 Solve Prob. 17.30, assuming that the force **P** is replaced by an 8-lb weight attached to cord C.

17.32 In the gear arrangement shown, gears A and C are attached to the rod ABC, which is free to rotate about B, while the inner gear B is fixed. Knowing that the system is at rest, determine the moment of the couple **M** which must be applied to rod ABC, if 3 s later the angular velocity of the rod is to be 300 rpm clockwise. Gears A and C weigh 3 lb each and may be considered as disks of radius 2 in.; rod ABC weighs 5 lb.

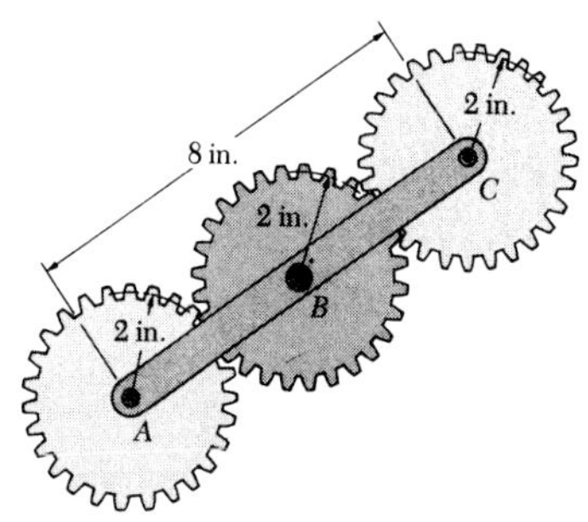

Fig. P17.32

Fig. P17.33

17.33 A 240-mm-diameter pipe of mass 16 kg rests on a 4-kg plate. The pipe and plate are initially at rest when a force **P** of magnitude 80 N is applied for 0.50 s. Knowing that $\mu_s = 0.25$ and $\mu_k = 0.20$ between the plate and both the pipe and the floor, determine (*a*) whether the pipe slides with respect to the plate, (*b*) the resulting velocities of the pipe and of the plate.

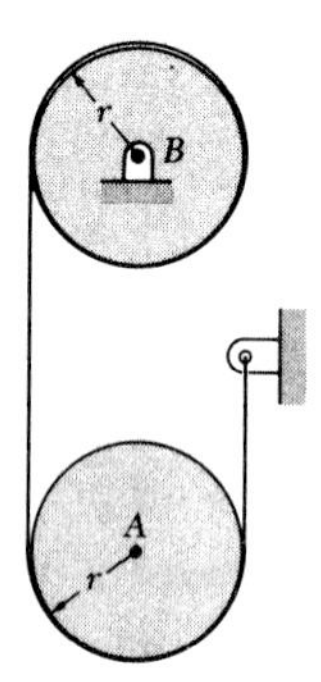

Fig. P17.34

17.34 Two uniform cylinders, each of weight $W = 14$ lb and radius $r = 5$ in., are connected by a belt as shown. If the system is released from rest, determine (*a*) the velocity of the center of cylinder *A* after 3 s, (*b*) the tension in the portion of belt connecting the two cylinders.

17.35 and 17.36 The 6-kg carriage is supported as shown by two uniform disks, each having a mass of 4 kg and a radius of 75 mm. Knowing that the carriage is initially at rest, determine the velocity of the carriage 2.5 s after the 10-N force has been applied. Assume that the disks roll without sliding.

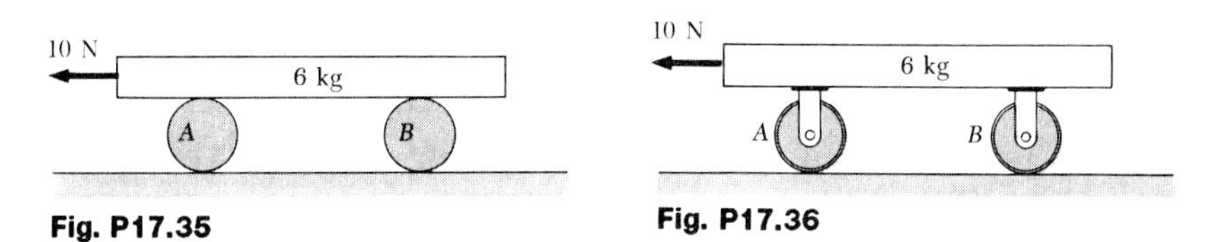

Fig. P17.35 Fig. P17.36

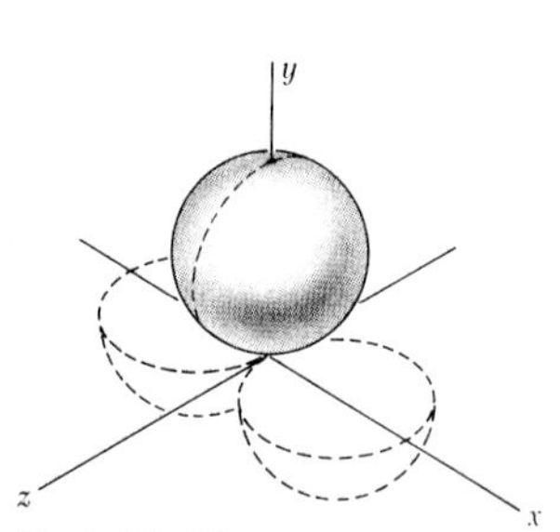

Fig. P17.37

17.37 A thin-walled spherical shell of mass *m* and radius *r* is used as a protective cover for a space shuttle experiment. In order to remove the cover, it is separated into two hemispherical shells which move to the final position shown. Knowing that during a test of the cover the initial angular velocity of the shell is $\boldsymbol{\omega} = \omega_0 \mathbf{j}$, determine the angular velocity of the cover after it has reached its final position. (*Hint.* For a thin-walled shell, $I_{\text{diameter}} = \frac{2}{3}mr^2$.)

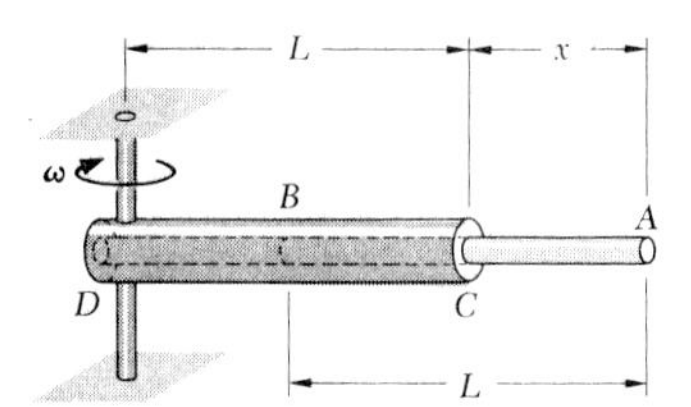

Fig. P17.38

17.38 The rod *AB* is of mass *m* and slides freely inside the tube *CD* which is also of mass *m*. The angular velocity of the assembly was ω_1 when the rod was entirely inside the tube ($x = 0$). Neglecting the effect of friction, determine the angular velocity of the assembly when $x = \frac{2}{3}L$.

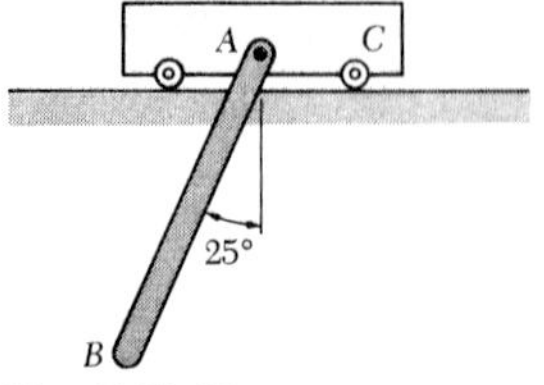

Fig. P17.39

17.39 The uniform rod *AB*, of weight 20 lb and length 3 ft, is attached to the 30-lb cart *C*. Knowing that the system is released from rest in the position shown and neglecting friction, determine (*a*) the velocity of point *B* as the rod *AB* passes through a vertical position, (*b*) the corresponding velocity of the cart *C*.

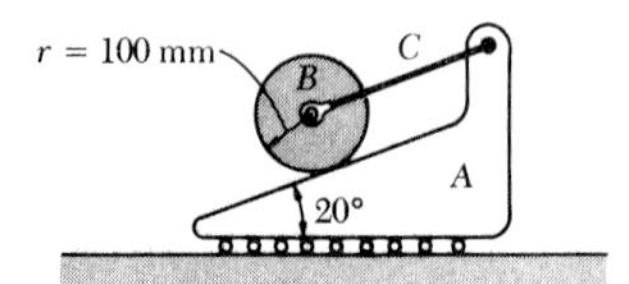

Fig. P17.40

17.40 The 3-kg cylinder *B* and the 2-kg wedge *A* are at rest in the position shown. The cord *C* connecting the cylinder and the wedge is then cut and the cylinder rolls without sliding on the wedge. Neglecting friction between the wedge and the ground, determine (*a*) the angular velocity of the cylinder after it has rolled 150 mm down the wedge, (*b*) the corresponding velocity of the wedge.

17.41 A 0.10-lb bullet is fired with a horizontal velocity of 1200 ft/s into a 20-lb wooden disk suspended from a pin support at A. Knowing that the disk is initially at rest, determine (a) the required distance h if the impulsive reaction at A is to be zero, (b) the corresponding velocity of the center G of the disk immediately after the bullet becomes embedded.

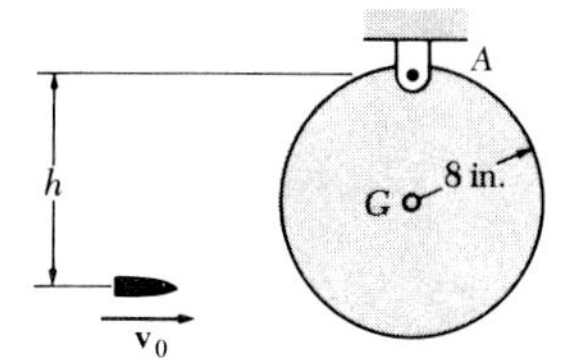

Fig. P17.41 and P17.42

17.42 A 0.10-lb bullet is fired with a horizontal velocity of 1200 ft/s into a 20-lb wooden disk suspended from a pin support at A. Knowing that $h = 10$ in. and that the disk is initially at rest, determine (a) the velocity of the center G of the disk immediately after the bullet becomes embedded, (b) the impulsive reaction at A, assuming that the bullet becomes embedded in 0.001 s.

17.43 An 8-kg wooden panel is suspended from a pin support at A and is initially at rest. A 2-kg metal sphere is released from rest at B and falls into a hemispherical cup C attached to the panel at a point located on its top edge. Assuming that the impact is perfectly plastic, determine the velocity of the mass center G of the panel immediately after the impact.

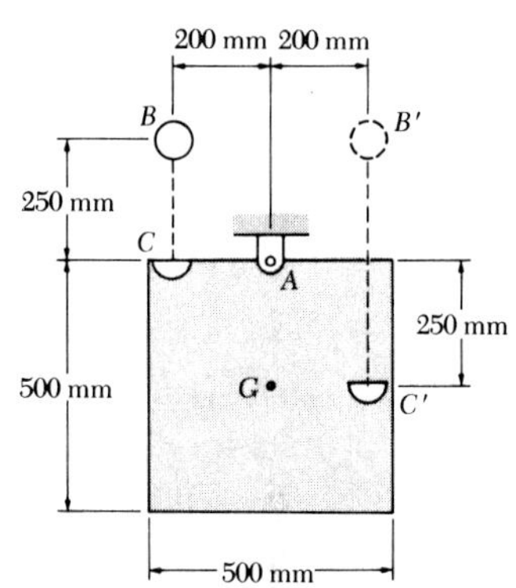

Fig. P17.43 and P17.44

17.44 An 8-kg wooden panel is suspended from a pin support at A and is initially at rest. A 2-kg metal sphere is released from rest at B' and falls into a hemispherical cup C' attached to the panel at the same level as the mass center G. Assuming that the impact is perfectly plastic, determine the velocity of the mass center G of the panel immediately after the impact.

17.45 A slender rod of length L is falling with a velocity $\mathbf{v}_1$ at the instant when the cords simultaneously become taut. Assuming that the impacts are perfectly plastic, determine the angular velocity of the rod and the velocity of its mass center immediately after the cords become taut.

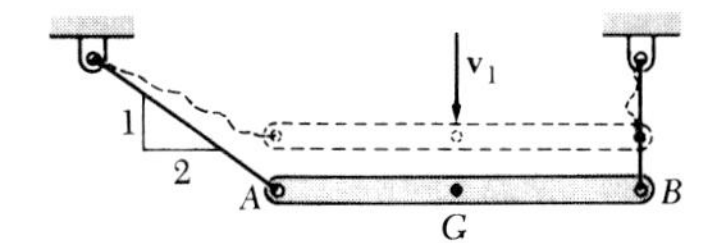

Fig. P17.45

17.46 A uniform square panel of side L and mass m is supported by a frictionless horizontal table. The panel is moving to the right with a velocity $\mathbf{v}_1$ when a hook located at the midpoint E of side AB engages the fixed pin F. Assuming that the impact is perfectly plastic, determine the angular velocity of the panel in its subsequent rotation about F.

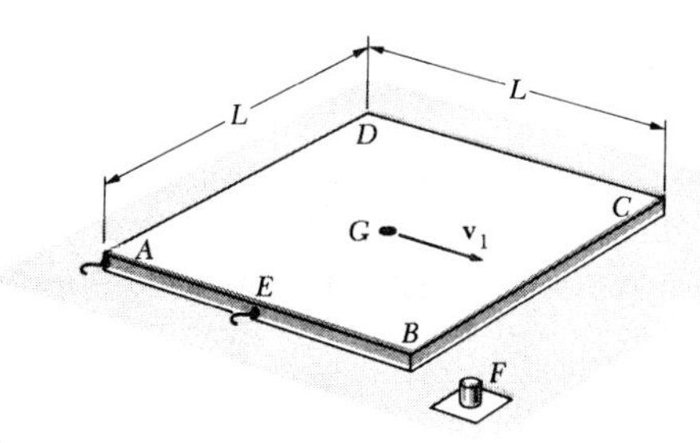

Fig. P17.46

17.47 A uniform slender rod AB is equipped at both ends with the hooks shown and is supported by a frictionless horizontal table. Initially the rod is hooked at A to a fixed pin C about which it rotates with the constant angular velocity ω_1. Suddenly end B of the rod hits and gets hooked to the pin D, causing end A to be released. Determine the magnitude of the angular velocity ω_2 of the rod in its subsequent rotation about D.

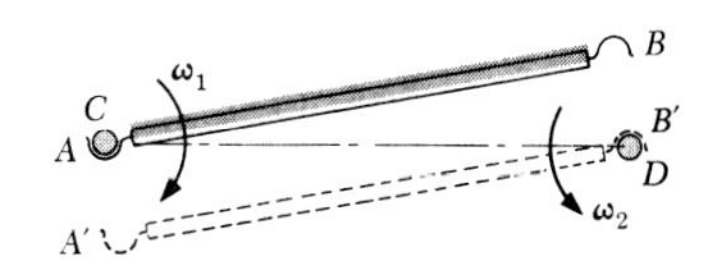

Fig. P17.47

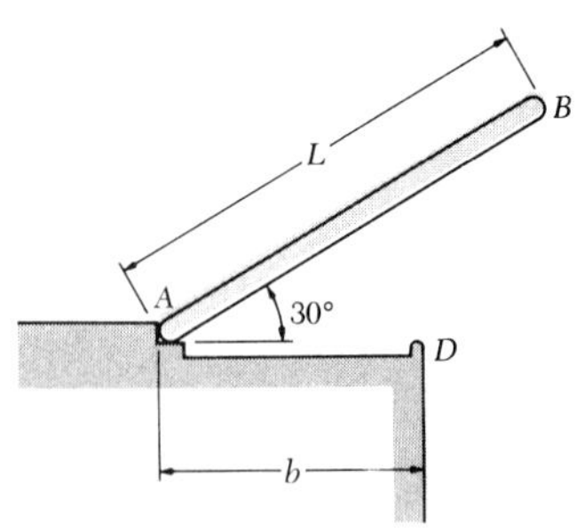

Fig. P17.48 and P17.49

17.48 A slender rod of length L and mass m is released from rest in the position shown and hits the edge D. Assuming perfectly plastic impact at D, determine for $b = 0.6L$, (*a*) the angular velocity of the rod immediately after the impact, (*b*) the maximum angle through which the rod will rotate after the impact.

17.49 A slender rod of mass m and length L is released from rest in the position shown and hits the edge D. Assuming perfectly elastic impact ($e = 1$) at D, determine the distance b for which the rod will rebound with no angular velocity.

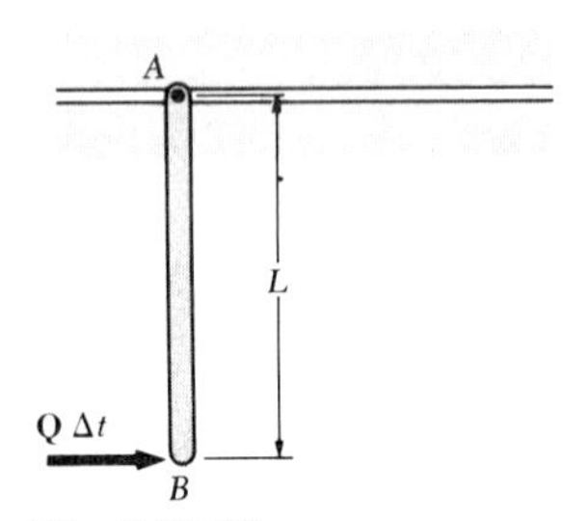

Fig. P17.50

17.50 Rod AB, of mass 1.5 kg and length $L = 0.5$ m, is suspended from a pin A which may move freely along a horizontal guide. If an impulse $\mathbf{Q}\,\Delta t = (1.8\ \text{N}\cdot\text{s})\mathbf{i}$ is applied at B, determine the maximum angle θ_m through which the rod will rotate during its subsequent motion.

17.51 For the rod of Prob. 17.50 , determine the magnitude $Q\,\Delta t$ of the impulse for which the maximum angle through which the rod will rotate is (*a*) 90°, (*b*) 120°.

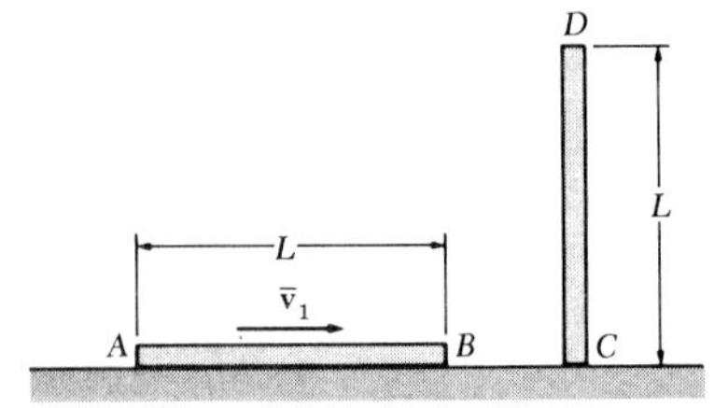

Fig. P17.52

17.52 A slender rod CD, of length L and mass m, is placed upright as shown on a frictionless horizontal surface. A second and identical rod AB is moving along the surface with a velocity $\bar{\mathbf{v}}_1$ when its end B strikes squarely end C of rod CD. Assuming that the impact is perfectly elastic, determine immediately after the impact, (*a*) the velocity of rod AB, (*b*) the angular velocity of rod CD, (*c*) the velocity of the mass center of rod CD.

17.53 Solve Prob. 17.52 , assuming $e = 0.50$.

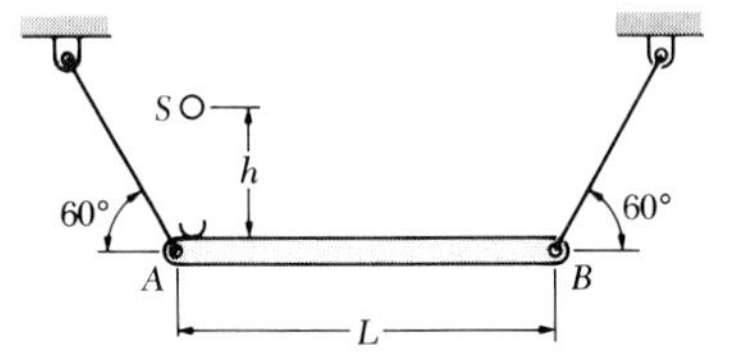

Fig. P17.54

17.54 A sphere of weight W_S is dropped from a height h and strikes at A the uniform slender plank AB of weight W which is held by two inextensible cords. Knowing that the impact is perfectly plastic and that the sphere remains attached to the plank, determine the velocity of the sphere immediately after impact.

17.55 Solve Prob. 17.54 when $W = 6$ lb, $W_S = 3$ lb, $L = 3$ ft, and $h = 4$ ft.

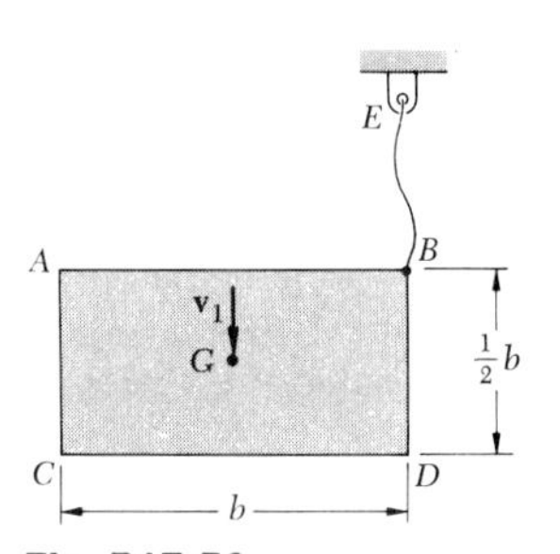

Fig. P17.56

17.56 The uniform plate $ABCD$ is falling with a velocity $\mathbf{v}_1$ when wire BE becomes taut. Assuming that the impact is perfectly plastic, determine the angular velocity of the plate and the velocity of its mass center immediately after the impact.

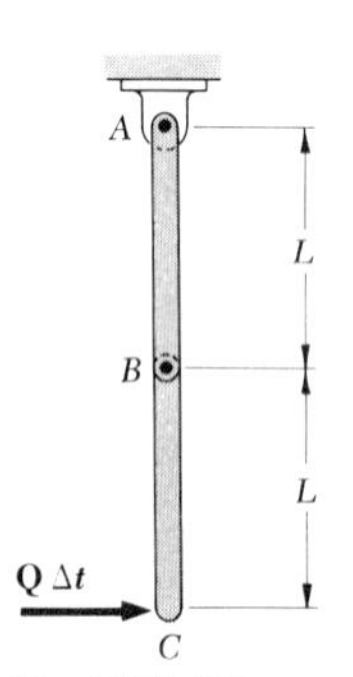

Fig. P17.57

17.57 Each of the bars AB and BC is of length $L = 15$ in. and weight $W = 2.5$ lb. Determine the angular velocity of each bar immediately after the impulse $\mathbf{Q}\,\Delta t = (0.30\ \text{lb}\cdot\text{s})\mathbf{i}$ is applied at C.

CHAPTER 17 COMPUTER PROBLEMS

17.C1 The 5-kg slender rod AB is welded to the 3-kg uniform disk which rotates about a pivot at A. A spring of constant 75 N/m is attached to the disk and is unstretched when rod AB is horizontal. The assembly is released from rest in the position shown. Determine and plot the angular velocity of the assembly as a function of the angle of rotation, starting from the initial rest position and ending at the angle of maximum rotation.

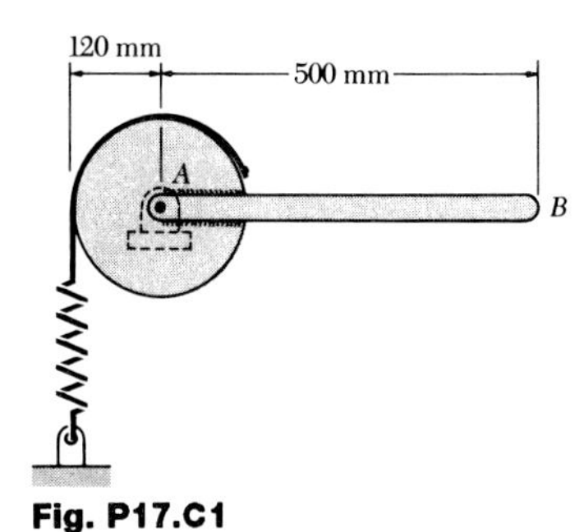

Fig. P17.C1

17.C2 Two 6-lb slender rods are welded to the edge of an 8-lb uniform disk as shown. The assembly is released from rest in the position shown and swings freely about the pivot C. Determine and plot the angular velocity of the assembly as a function of the angle of rotation. Find the maximum velocity attained by the assembly.

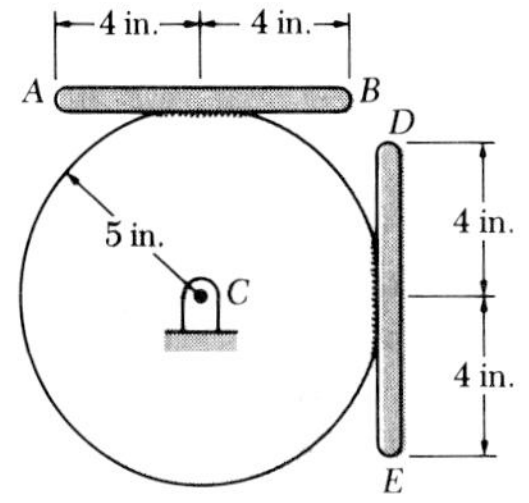

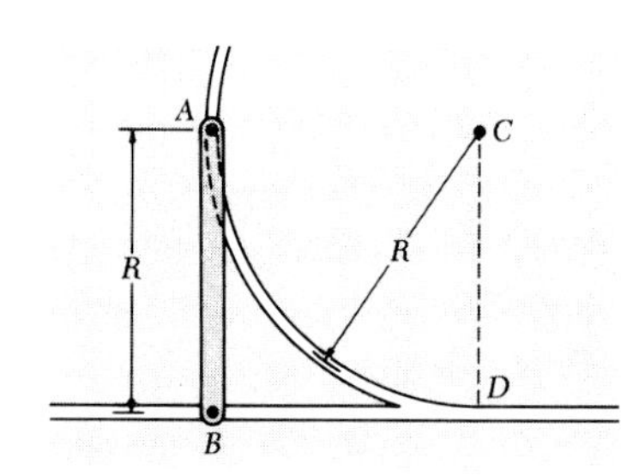

Fig. P17.C2

17.C3 The motion of a slender rod of length $R = 2$ m is guided by pins at A and B which slide freely in slots cut in a vertical plate as shown. The rod is set in motion when the end B is moved slightly to the left and then released. Determine and plot the angular velocity of the rod and the velocity of its mass center as functions of the height of end A starting from the position shown and ending when the end A reaches point D.

Fig. P17.C3

17.C4 A tape moves over the two drums shown. Drum A weighs 0.90 lb and has a radius of gyration of 0.70 in., while drum B weighs 1.8 lb and has a radius of gyration of 1.10 in. In the portion of the tape above drum A the tension is constant and equal to $T_A = 0.5$ lb. The tape is initially at rest when a constant tensile force T_B is applied. Determine and plot the velocity of the tape after 0.2 s for values of T_B from zero to 5 lb. Determined and plot the tension in the portion of the tape between the drums as a function of T_B.

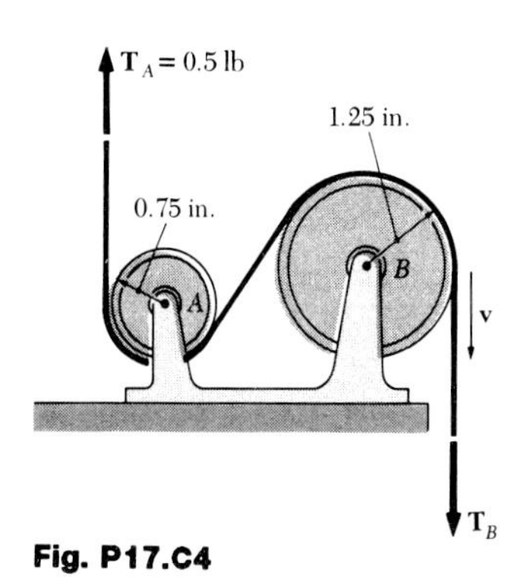

Fig. P17.C4

17.C5 A 1.5-kg tube AB may slide freely on rod DE which in turn may rotate freely in a horizontal plane. Initially the assembly is rotating with an angular velocity $\omega = 6$ rad/s and the tube is held in position by a cord. The moment of inertia of the rod and bracket about the vertical axis of rotation is 0.20 kg-m^2 and the centroidal moment of inertia of the tube about a vertical axis is 0.0015 kg-m^2. If the cord suddenly breaks, determine and plot the angular velocity of the assembly as a function of the position of the tube until it moves to the end E.

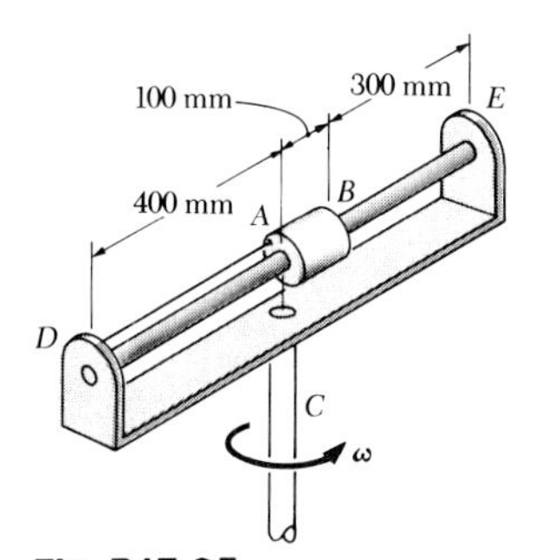

Fig. P17.C5

CHAPTER 18
KINETICS OF RIGID BODIES IN THREE DIMENSIONS

SECTION 18.1 to 18.4

18.1 A thin homogeneous square plate of mass m and side a is welded to a vertical shaft AB with which it forms an angle of 45°. Knowing that the shaft rotates with an angular velocity $\boldsymbol{\omega}$, determine the angular momentum of the plate about A.

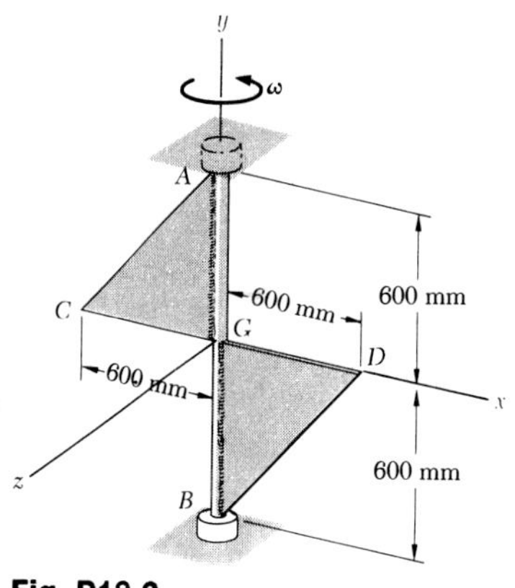

Fig. P18.2

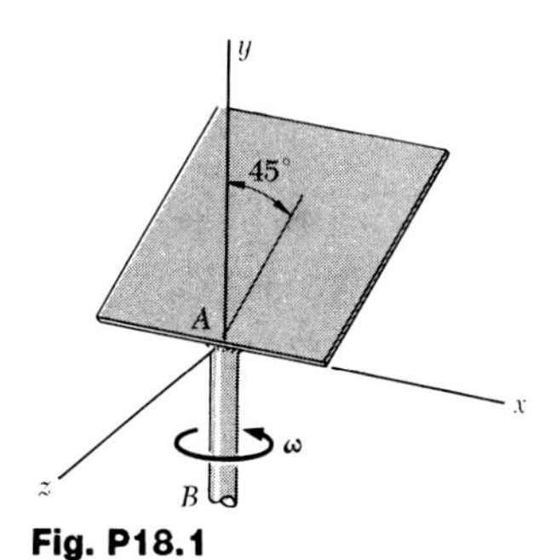

Fig. P18.1

18.2 Two triangular plates, each of mass 5 kg, are welded to a vertical shaft AB. Knowing that the system rotates at the constant rate $\omega = 8$ rad/s, determine its angular momentum about G.

18.3 Two L-shaped arms, each weighing 4 lb, are welded at the third points of the 2-ft shaft AB. Knowing that shaft AB rotates at the constant rate $\omega = 240$ rpm, determine (*a*) the angular momentum of the body about A, (*b*) the angle formed by the angular momentum and the shaft AB.

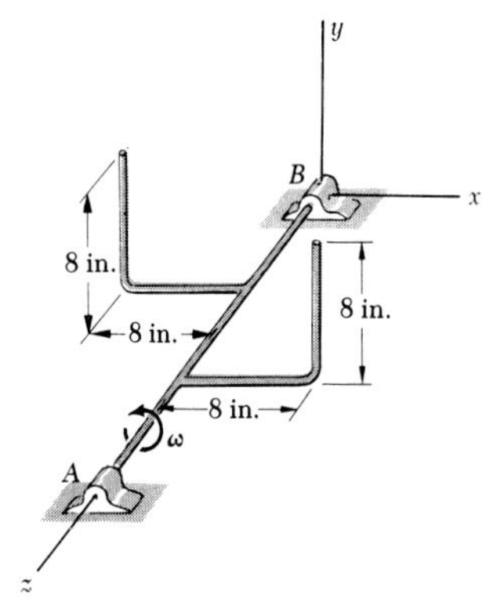

Fig. P18.3

18.4 For the body of Prob. 18.3, determine (*a*) the angular momentum about B, (*b*) the angle formed by the angular momentum and the shaft BA.

18.5 and 18.6 Two uniform rods AB and CD are welded together at B to form a T-shaped assembly of total mass m. The assembly is suspended from a ball-and-socket joint at A and is hit at C in a direction perpendicular to its plane (in the negative z direction). Denoting the corresponding impulse by $\mathbf{F}\,\Delta t$, determine immediately after the impact (*a*) the angular velocity of the assembly, (*b*) its instantaneous axis of rotation.

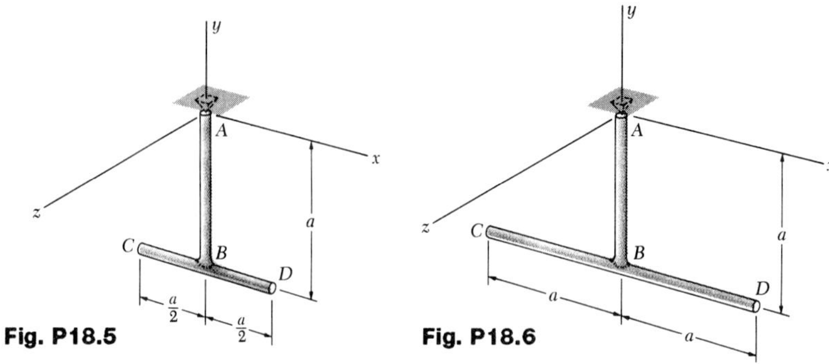

Fig. P18.5

Fig. P18.6

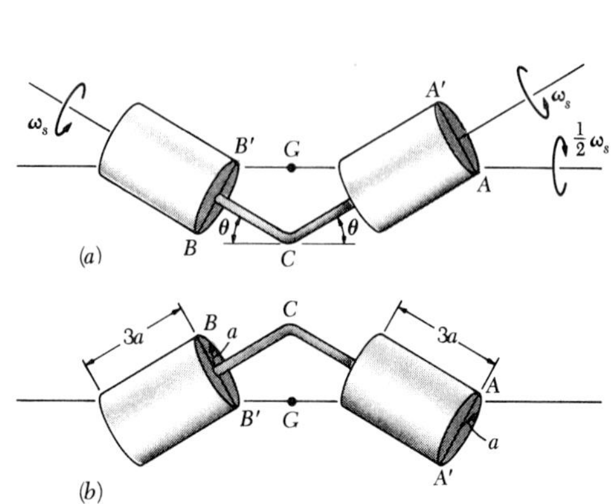

Fig. P18.7

18.7 In order to analyze the twist performed by cats to land on their paws after being dropped upside down, the model shown may be used. It consists of two homogeneous cylinders, each of mass m, radius a, and length $3a$, mounted on a bent shaft. Small motors located inside the cylinders make it possible to spin each cylinder about the shaft at the rate ω_s after the assembly has been released from rest in position a. Observing that the entire assembly must rotate at the rate $\frac{1}{2}\omega_s$ about a horizontal axis through its mass center G if it is to complete a half turn about that axis when each cylinder has spun through 360° with respect to the shaft (position b), determine the required angle θ that each portion of shaft should form with the horizontal. Neglect the mass of the shaft and of the motors. (*Hint.* Since no external force except its weight acts on the assembly, the angular momentum of the assembly about G must remain equal to zero).

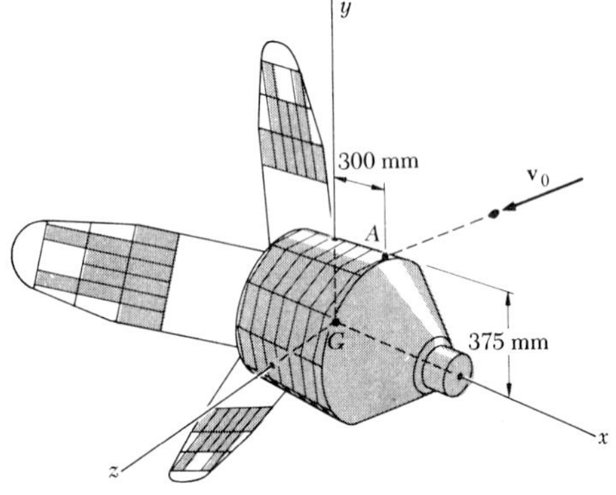

Fig. P18.8

18.8 A satellite of mass 160 kg has no angular velocity when it is struck at A by a 20-g meteorite traveling with a velocity $\mathbf{v}_0 = -(720 \text{ m/s})\mathbf{i} - (540 \text{ m/s})\mathbf{j} + (1200 \text{ m/s})\mathbf{k}$ relative to the satellite. Knowing that the radii of gyration of the satellite are $\bar{k}_x = 300$ mm and $\bar{k}_y = \bar{k}_z = 400$ mm, determine the angular velocity of the satellite immediately after the meteorite has become imbedded.

18.9 Solve Prob. 18.8 , assuming that the satellite was initially spinning about its axis of symmetry with an angular velocity of 1.25 rad/s clockwise as viewed from the positive x axis.

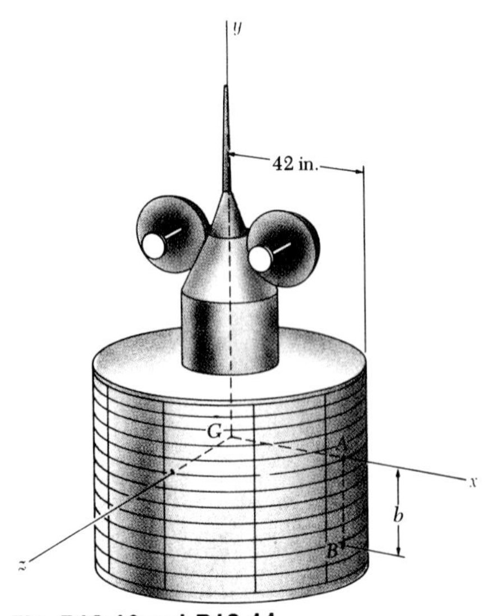

Fig. P18.10 and P18.11

18.10 An 800-lb geostationary satellite is spinning with the angular velocity $\boldsymbol{\omega}_0 = (1.5 \text{ rad/s})\mathbf{j}$ when it is hit at A by a 6-oz meteorite traveling at a speed $v_0 = 4500$ ft/s relative to the satellite. Knowing that the radii of gyration of the satellite are $\bar{k}_x = \bar{k}_z = 28.8$ in. and $\bar{k}_y = 32.4$ in., and that immediately after the meteorite has become imbedded the angular velocity of the satellite is observed to be $\boldsymbol{\omega} = (0.6 \text{ rad/s})\mathbf{j} + (0.37 \text{ rad/s})\mathbf{k}$, determine the components of the velocity $\mathbf{v}_0$ of the meteorite relative to the satellite.

18.11 An 800-lb geostationary satellite is spinning with the angular velocity $\boldsymbol{\omega}_0 = (1.5 \text{ rad/s})\mathbf{j}$ when it is hit at B by a 6-oz meteorite traveling with the velocity $\mathbf{v}_0 = -(2800 \text{ ft/s})\mathbf{i} + (2900 \text{ ft/s})\mathbf{j} - (2000 \text{ ft/s})\mathbf{k}$ relative to the satellite. Knowing that the radii of gyration of the satellite are $\bar{k}_x = \bar{k}_z = 28.8$ in. and $\bar{k}_y = 32.4$ in., and that immediately after the meteorite has become imbedded the x component of the angular velocity $\boldsymbol{\omega}$ of the satellite is observed to be $\omega_x = 0.3$ rad/s, determine (*a*) the distance b, (*b*) the y and z components of $\boldsymbol{\omega}$.

18.12 Determine the kinetic energy of the plate of Prob. 18.1.

18.13 Determine the kinetic energy of the body of Prob. 18.3 .

18.14 Determine the kinetic energy of the satellite of Prob. 18.8 in its motion about its mass center after the collision with the meteorite.

18.15 Determine the change in the kinetic energy of the satellite of Prob. 18.10 in its motion about its mass center due to the collision with the meteorite.

18.16 Gear A rolls on the fixed gear B and rotates about the axle AD of length $L = 500$ mm which is rigidly attached at D to the vertical shaft DE. The shaft DE is made to rotate with a constant angular velocity $\boldsymbol{\omega}_1$ of magnitude 4 rad/s. Assuming that gear A can be approximated by a thin disk of mass 2 kg and radius $a = 100$ mm, and that $\beta = 30°$, determine (a) the angular momentum of gear A about point D, (b) the kinetic energy of gear A.

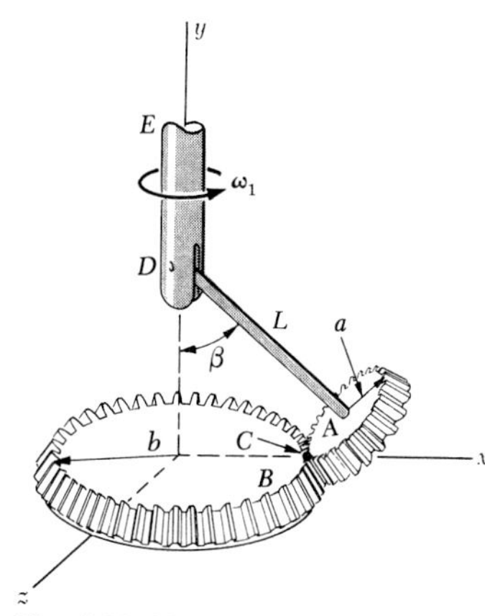

Fig. P18.16

SECTIONS 18.5 to 18.8

18.17 Determine the rate of change $\dot{\mathbf{H}}_A$ of the angular momentum $\mathbf{H}_A$ of the plate of Prob. 18.1, knowing that its angular velocity $\boldsymbol{\omega}$ remains constant.

18.18 Determine the rate of change $\dot{\mathbf{H}}_A$ of the angular momentum $\mathbf{H}_A$ of the plate of Prob. 18.1, knowing that it has an angular velocity $\boldsymbol{\omega} = \omega\mathbf{j}$ and an angular acceleration $\boldsymbol{\alpha} = \alpha\mathbf{j}$.

18.19 A thin homogeneous rod AB of mass m and length $2b$ is welded at its midpoint G to a vertical shaft GD. Knowing that the shaft rotates with a constant angular velocity $\boldsymbol{\omega}$, determine the couple exerted by the shaft on rod AB.

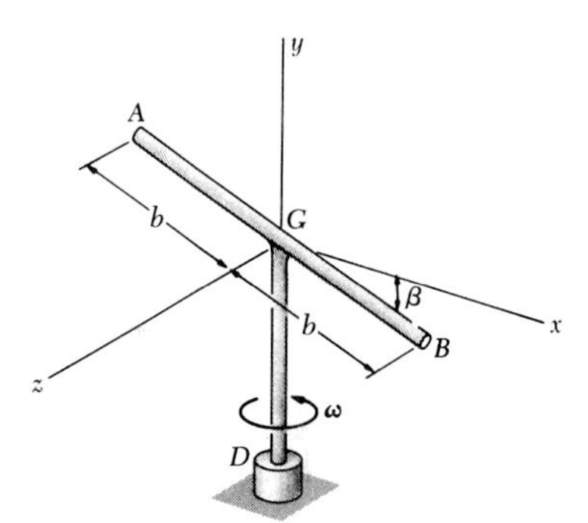

Fig. P18.19

18.20 A thin homogeneous rod of weight w per unit length is used to form the shaft shown. Knowing that the shaft rotates with a constant angular velocity $\boldsymbol{\omega}$, determine the dynamic reactions at A and B.

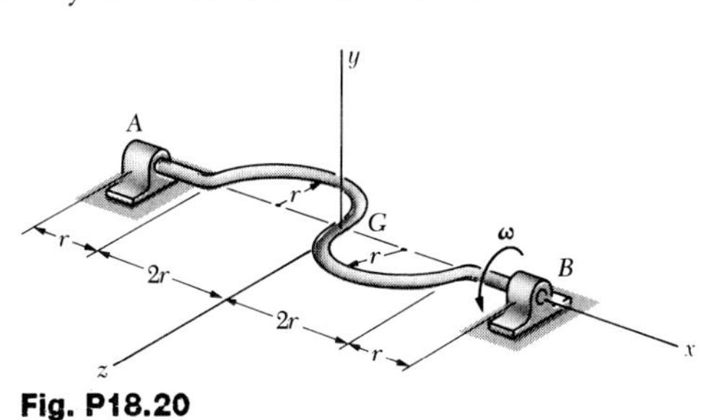

Fig. P18.20

18.21 Two L-shaped arms, each weighing 4 lb, are welded at the third points of a 2-ft horizontal shaft supported by bearings at A and B. Knowing that the shaft rotates at the constant rate $\omega = 240$ rpm, determine the dynamic reactions at A and B.

18.22 Two L-shaped arms, each weighing 4 lb, are welded at the third points of a 2-ft horizontal shaft supported by bearings at A and B. The assembly is a rest ($\omega = 0$) when a couple of moment $\mathbf{M}_0 = (20\text{ lb}\cdot\text{in.})\mathbf{k}$ is applied to the shaft. Determine (a) the resulting angular acceleration of the shaft, (b) the dynamic reactions at A and B immediately after the couple has been applied.

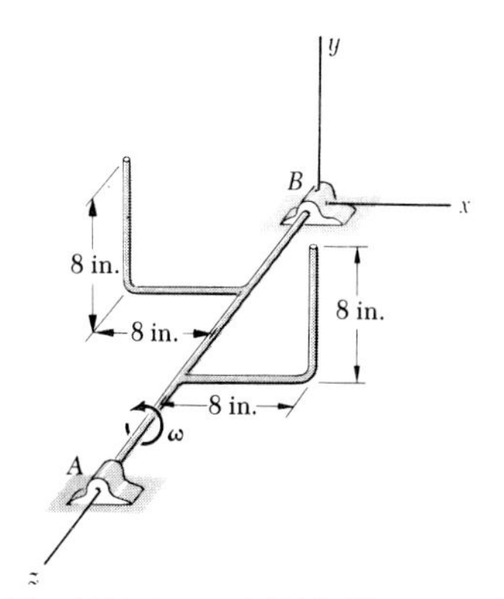

Fig. P18.21 and P18.22

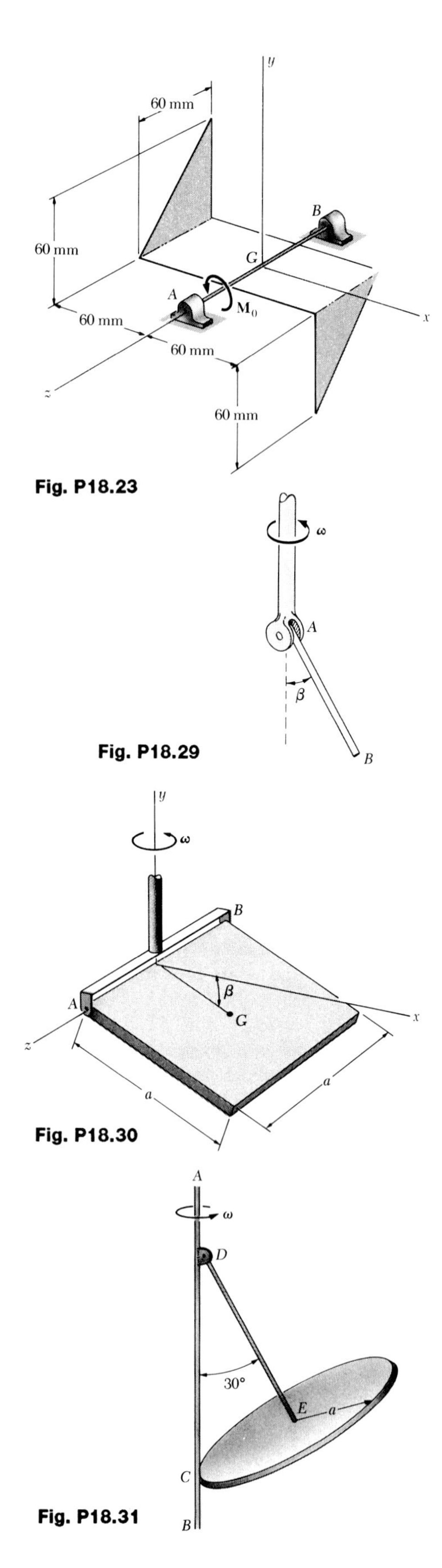

Fig. P18.23

Fig. P18.29

Fig. P18.30

Fig. P18.31

18.23 The sheet-metal component shown is of uniform thickness and has a mass of 400 g. It is attached to a light axle supported by bearings at A and B located 120 mm apart. The component is at rest when it is subjected to a couple $\mathbf{M}_0$ as shown. If the resulting angular acceleration is $\boldsymbol{\alpha} = (12 \text{ rad/s}^2)\mathbf{k}$, determine ($a$) the couple $\mathbf{M}_0$, (b) the dynamic reactions at A and B immediately after the couple has been applied.

18.24 Knowing that the shaft of Prob. 18.20 is originally at rest ($\omega = 0$), determine (a) the couple $\mathbf{M}_0$ required to impart to the shaft an acceleration $\boldsymbol{\alpha} = \alpha\mathbf{i}$, ($b$) the dynamic reactions at A and B immediately after the couple $\mathbf{M}_0$ has been applied.

18.25 For the sheet-metal component of Prob. 18.23, determine (a) the angular velocity of the component 0.5 s after the couple $\mathbf{M}_0$ has been applied to it, (b) the dynamic reactions at A and B at that time.

18.26 Knowing that the shaft GD in Prob. 18.19 has an angular velocity $\boldsymbol{\omega} = \omega\mathbf{j}$ and an angular acceleration $\boldsymbol{\alpha} = \alpha\mathbf{j}$ at the instant shown, determine the components of the couple exerted by the shaft on rod AB.

18.27 Each wheel of an automobile weighs 45 lb and has a radius of gyration of 10 in. The automobile travels around an unbanked curve of radius 600 ft at a speed of 60 mi/h. Knowing that each wheel is of 24-in. diameter and that the transverse distance between the wheels is 58 in., determine the additional normal reaction at each outside wheel due to the rotation of the car.

18.28 An airplane has a single four-bladed propeller which weighs 250 lb and has a radius of gyration of 32 in. Knowing that the propeller rotates at 1800 rpm clockwise as seen from the front, determine the magnitude of the couple exerted by the propeller on its shaft and the resulting effect on the airplane when the pilot executes a horizontal turn of 1500-ft radius to the left at a speed of 375 mi/h.

18.29 A uniform rod AB of length l and mass m is attached to the pin of a clevis which rotates with a constant angular velocity $\boldsymbol{\omega}$. Determine (a) the constant angle β that the rod forms with the vertical, (b) the range of values of ω for which the rod remains vertical ($\beta = 0$).

18.30 A uniform square plate of side a and mass m is hinged at A and B to a clevis which rotates with a constant angular velocity $\boldsymbol{\omega}$. Determine (a) the value of ω for which the angle β maintains the constant value $\beta = 30°$, (b) the range of values of ω for which the plate remains vertical ($\beta = 90°$).

18.31 A disk of mass m and radius a is rigidly attached to a rod DE of negligible mass. The rod DE is attached to a vertical shaft AB by a clevis at D and the disk leans against the shaft at C. Noting that when the shaft AB is made to rotate, the same point of the disk will remain in contact with the shaft at C, determine the magnitude of the angular velocity $\boldsymbol{\omega}$ for which the reaction at C will be zero.

18.32 Two disks, each of weight 10 lb, and radius 12 in., spin as shown at 1200 rpm about the rod AB, which is attached to shaft CD. The entire system is made to rotate about the z axis with an angular velocity $\mathbf{\Omega}$ of 60 rpm. (*a*) Determine the dynamic reactions at C and D due to gyroscopic action as the system passes through the position shown. (*b*) Solve part *a* assuming that the direction of spin of disk B is reversed.

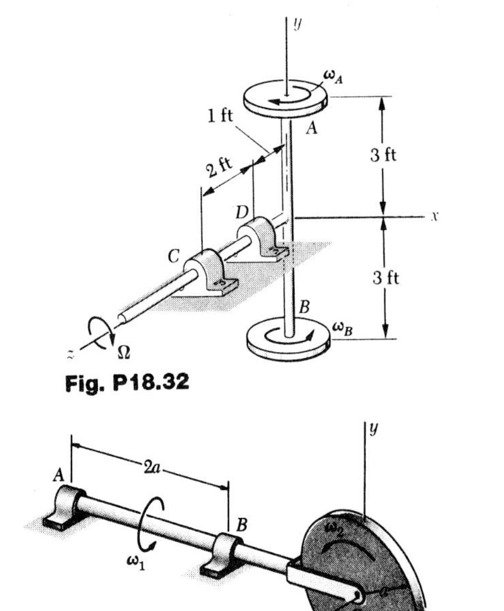

Fig. P18.32

18.33 A thin homogeneous disk of mass m and radius a is held by a fork-ended horizontal rod ABC. The disk and the rod rotate with the angular velocities shown. Assuming that both ω_1 and ω_2 are constant, determine the dynamic reactions at A and B.

Fig. P18.33

18.34 A thin disk of mass $m = 5$ kg rotates with an angular velocity $\boldsymbol{\omega}_2$ with respect to the bent axle ABC, which itself rotates with an angular velocity $\boldsymbol{\omega}_1$ about the y axis. Knowing that $\omega_1 = 3$ rad/s and $\omega_2 = 8$ rad/s and that both are constant, determine the force-couple system representing the dynamic reaction at the support at A.

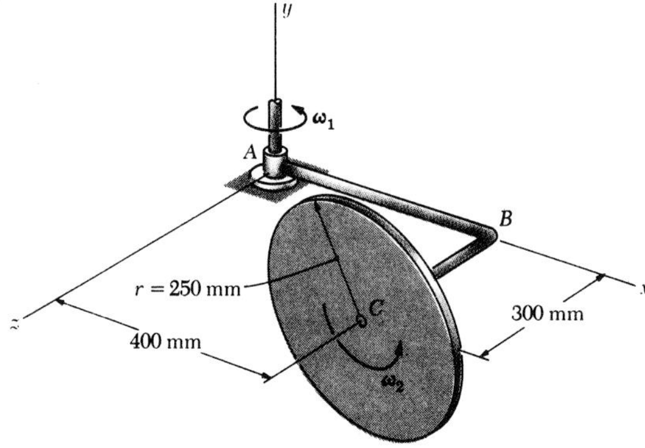

Fig. P18.34

18.35 For the disk of Prob. 18.34, determine (*a*) the couple $M_1\mathbf{j}$ which should be applied to the bent axle ABC to give it an angular acceleration $\boldsymbol{\alpha}_1 = (6 \text{ rad/s}^2)\mathbf{j}$ when $\omega_1 = 3$ rad/s, knowing that the disk rotates at the constant rate $\omega_2 = 8$ rad/s, (*b*) the force-couple system representing the dynamic reaction at A at that instant. Assume that the bent axle ABC has a negligible mass.

18.36 For the disk of Prob. 18.34, determine (*a*) the couple $M_1\mathbf{j}$ which should be applied to the bent axle ABC to keep it rotating at the constant rate $\omega_1 = 3$ rad/s at an instant when the angular velocity $\boldsymbol{\omega}_2$ of the disk has a magnitude of 8 rad/s and decreases at the rate of 2 rad/s^2 due to the axle friction at C, (*b*) the force-couple representing the dynamic reaction at A at that instant.

18.37 For the disk of Prob. 18.33, determine (*a*) the couple $M_1\mathbf{i}$ which should be applied to rod AB to give it an angular acceleration $\alpha_1\mathbf{i}$ when the rod has an angular velocity $\omega_1\mathbf{i}$, knowing that the disk rotates at the constant rate ω_2, (*b*) the dynamic reactions at A and B at that instant.

18.38 The slender homogeneous rod AB of mass m and length L is made to rotate at the constant rate ω_2 about the horizontal z axis, while the frame CD is made to rotate at the constant rate ω_1 about the vertical y axis. Express as a function of the angle θ (*a*) the couple $\mathbf{M}_1$ required to maintain the rotation of the frame, (*b*) the couple $\mathbf{M}_2$ required to maintain the rotation of the rod, (*c*) the dynamic reactions at the supports C and D.

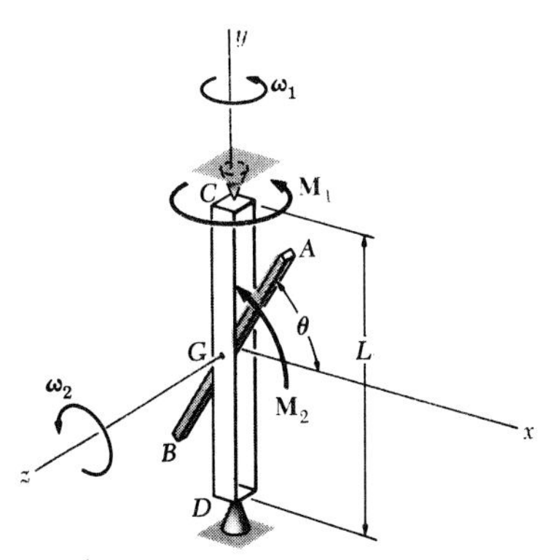

Fig. P18.38

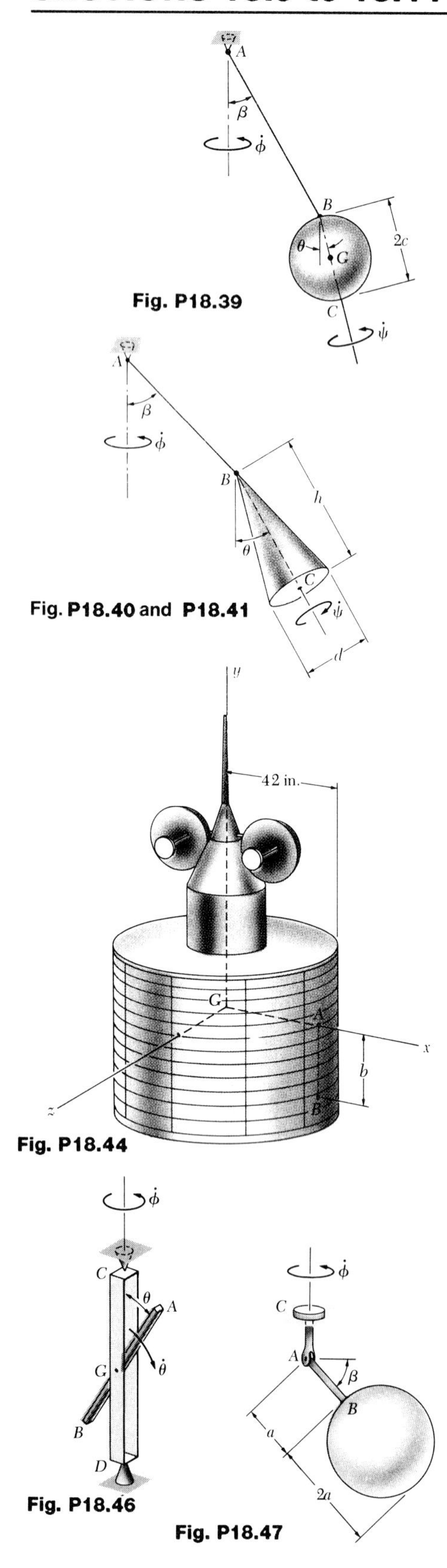

Fig. P18.39

Fig. P18.40 and P18.41

Fig. P18.44

Fig. P18.46

Fig. P18.47

18.39 A homogeneous sphere of radius $c = 30$ mm is attached as shown to a cord AB. The cord forms an angle $\beta = 30°$ with the vertical and is observed to precess at the constant rate $\dot{\phi} = 6$ rad/s about the vertical through A. Determine the angle θ that the diameter BC forms with the vertical, knowing that the sphere (*a*) has no spin, (*b*) spins about its diameter BC at the rate $\dot{\psi} = 30$ rad/s, (*c*) spins about BC at the rate $\dot{\psi} = -30$ rad/s.

18.40 A homogeneous cone of height h and with a base of diameter $d < h$ is attached as shown to a cord AB. The cone spins about its axis BC at the constant rate $\dot{\psi}$ and precesses about the vertical through A at the constant rate $\dot{\phi}$. Determine the angle β for which the axis BC of the cone is aligned with the cord AB ($\theta = \beta$).

18.41 A homogeneous cone of height $h = 12$ in. and base diameter $d = 6$ in. is attached as shown to a cord AB. Knowing that the angles that the cord AB and the axis BC of the cone form with the vertical are, respectively, $\beta = 45°$ and $\theta = 30°$, and that the cone precesses at the constant rate $\dot{\phi} = 8$ rad/s in the sense indicated, determine (*a*) the rate of spin $\dot{\psi}$ of the cone about its axis BC, (*b*) the length of cord AB.

18.42 Determine the precession axis and the rates of precession and spin of the satellite of Prob. 18.8 after the impact, knowing that before the impact the satellite was spinning about its axis of symmetry with the angular velocity $\boldsymbol{\omega}_0 = -(1.25 \text{ rad/s})\mathbf{i}$.

18.43 Determine the precession axis and the rates of precession and spin of the satellite of Prob. 18.8 after the impact.

18.44 An 800-lb geostationary satellite is spinning with the angular velocity $\boldsymbol{\omega}_0 = (1.5 \text{ rad/s})\mathbf{j}$ when it is hit at B by a 6-oz meteorite traveling with the velocity $\mathbf{v}_0 = -(1600 \text{ ft/s})\mathbf{i} + (1300 \text{ ft/s})\mathbf{j} + (4000 \text{ ft/s})\mathbf{k}$ relative to the satellite. Knowing that $b = 20$ in. and that the radii of gyration of the satellite are $\bar{k}_x = \bar{k}_z = 28.8$ in. and $\bar{k}_y = 32.4$ in., determine the precession axis and the rates of precession and spin of the satellite after the impact.

18.45 Solve Prob. 18.44, assuming that the meteorite hits the satellite at A instead of B.

18.46 The slender homogeneous rod AB of mass m and length L is free to rotate about a horizontal axle through its mass center G. The axle is supported by a frame of negligible mass which is free to rotate about the vertical CD. Knowing that, initially, $\theta = \theta_0$, $\dot{\theta} = 0$, and $\dot{\phi} = \dot{\phi}_0$, show that the rod will oscillate about the horizontal axle and determine (*a*) the range of values of the angle θ during this motion, (*b*) the maximum value of $\dot{\theta}$, (*c*) the minimum value of $\dot{\phi}$.

18.47 A homogeneous sphere of mass m and radius a is welded to a rod AB of negligible mass, which is connected by a clevis to the vertical shaft AC. The rod and sphere can rotate freely about a horizontal axis at A, and the shaft AC can rotate freely about a vertical axis. The system is released in the position $\beta = 0$ with an angular velocity $\dot{\phi}_0$ about the vertical axis and no angular velocity about the horizontal axis. Knowing that the largest value of β in the ensuing motion is $30°$, determine (*a*) the initial angular velocity $\dot{\phi}_0$, (*b*) the value of $\dot{\phi}$ when $\beta = 30°$.

18.48 A homogeneous sphere of mass m and radius a is welded to a rod AB of negligible mass, which is held by a ball-and-socket support at A. The sphere is released in the position $\beta = 0$ with a rate of precession $\dot{\phi}_0 = \sqrt{17g/11a}$ and with no spin or nutation. Determine the largest value of β in the ensuing motion.

18.49 A homogeneous sphere of mass m and radius a is welded to a rod AB of length a and negligible mass, which is held by a ball-and-socket support at A. The sphere is released in the position $\beta = 0$ with a rate of precession $\dot{\phi} = \dot{\phi}_0$ and with no spin or nutation. Knowing that the largest value of β in the ensuing motion is 30°, determine (*a*) the rate of precession $\dot{\phi}_0$ of the sphere in its initial position, (*b*) the rates of precession and spin when $\beta = 30°$.

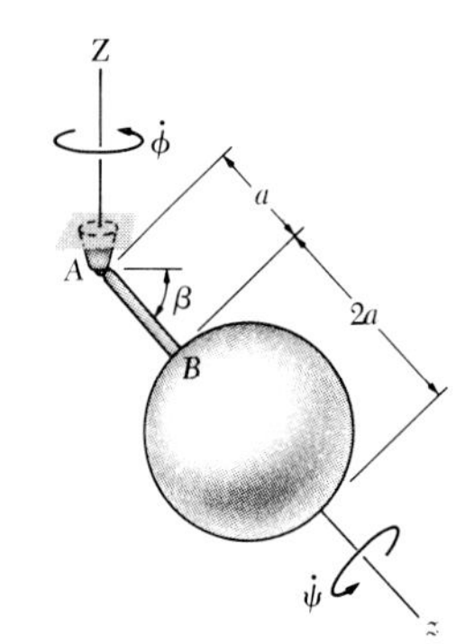

Fig. P18.48 and P18.49

18.50 A thin homogeneous disk of mass 600 g and radius 100 mm rotates at a constant rate $\omega_2 = 30$ rad/s with respect to the arm ABC, which itself rotates at a constant rate $\omega_1 = 15$ rad/s about the x axis. Determine the angular momentum of the disk about point C.

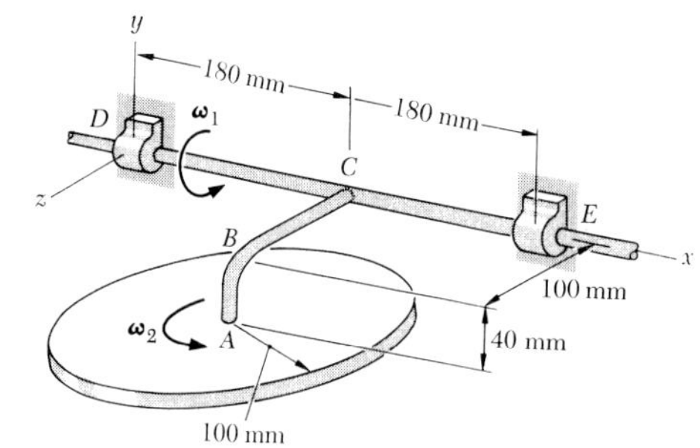

Fig. P18.50

18.51 Gear A rolls on the fixed gear B and rotates about the axle AD of length $L = 500$ mm which is rigidly attached at D to the vertical shaft DE. The shaft DE is made to rotate with a constant angular velocity $\boldsymbol{\omega}_1$ of magnitude 4 rad/s. Assuming that gear A can be approximated by a thin disk of mass 2 kg and radius $a = 100$ mm, and that $\beta = 30°$, determine (*a*) the angular momentum of gear A about point D, (*b*) the kinetic energy of gear A.

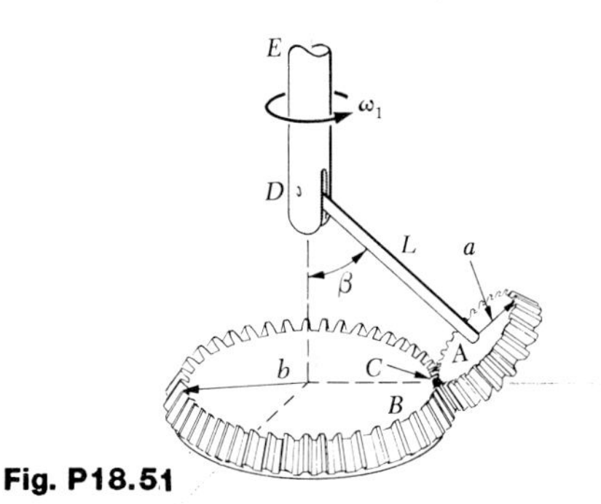

Fig. P18.51

18.52 A homogeneous sphere of radius a and mass m is attached to a light rod of length $4a$. The rod forms an angle of 30° with the vertical and rotates about AC at the constant rate $\Omega = \sqrt{g/a}$. (*a*) Assuming that the sphere does not spin about the rod ($\dot{\psi} = 0$), determine the tension in the cord BC and the kinetic energy of the sphere. (*b*) Determine the spin $\dot{\psi}$ (magnitude and sense) which should be given to the sphere if the tension in the cord BC is to be zero. What is the corresponding kinetic energy of the sphere?

18.53 The space station shown is known to precess about the fixed direction OC at the rate of one revolution per hour. Assuming that the station is dynamically equivalent to a homogeneous cylinder of length 30 m and radius 3 m, determine the rate of spin of the station about its axis of symmetry.

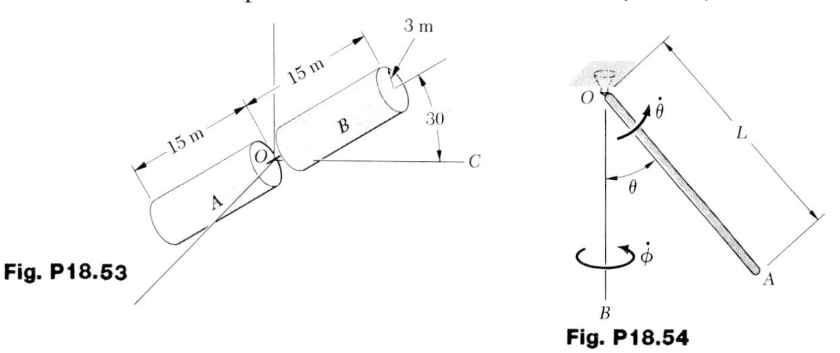

Fig. P18.53

Fig. P18.54

18.54 A slender homogeneous rod OA of mass m and length L is supported by a ball-and-socket joint at O and may swing freely under its own weight. If the rod is held in a horizontal position ($\theta = 90°$) and given an initial angular velocity $\dot{\phi}_0 = \sqrt{8g/L}$ about the vertical OB, determine (*a*) the smallest value of θ in the ensuing motion, (*b*) the corresponding value of the angular velocity $\dot{\phi}$ of the rod about OB. (*Hint.* Apply the principle of conservation of energy and the principle of impulse and momentum, observing that since $\Sigma M_{OB} = 0$, the component of $\mathbf{H}_O$ along OB must be constant.)

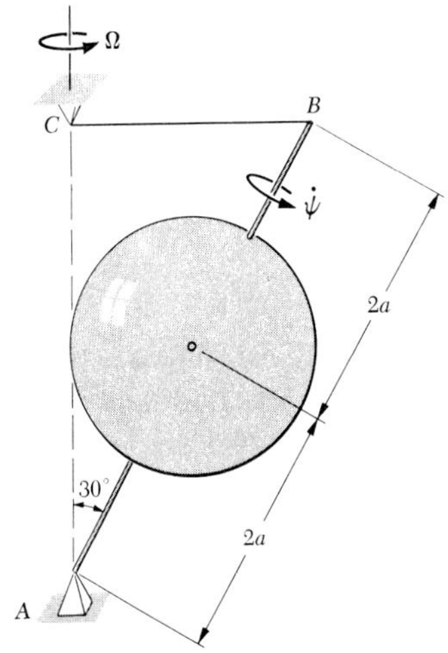

Fig. P18.52

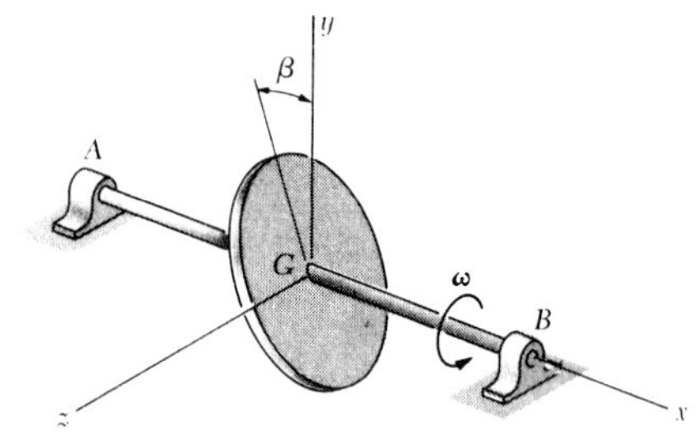

Fig. P18.C1

18.C1

A thin homogeneous disk of mass m and radius r is mounted on the horizontal axle AB. The plane of the disk forms an angle β with the plane normal to the axle. The axle rotates with a constant angular velocity ω. For values of β from zero to 90°, determine and plot (a) the angle θ formed by the axle and the angular momentum of the disk about G, (b) the kinetic energy of the disk, and (c) the magnitude of the rate of change of the angular momentum of the disk about G.

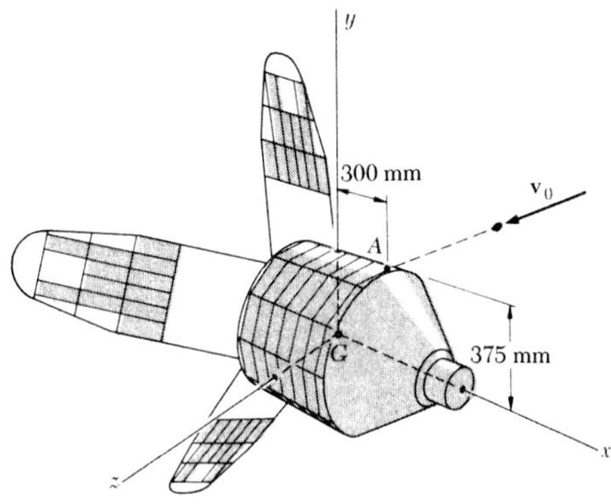

Fig. P18.C2

18.C2

A satellite of mass 160 kg has no angular velocity when it is struck at A by a 20-g meteorite traveling with a velocity $\mathbf{v}_0 = -(1500 \text{ m/s})\cos\theta\, \mathbf{i} - (1500 \text{ m/s})\sin\theta\, \mathbf{j}$ relative to the satellite. The radii of gyration of the satellite are $\bar{k}_x = 300$ mm and $\bar{k}_y = \bar{k}_z = 400$ mm. Determine and plot the angular velocity of the satellite immediately after the meteorite has become imbedded for values of θ from -90° to $+90^{\circ}$. Determine the maximum angular velocity ω_m and the corresponding angle θ_m.

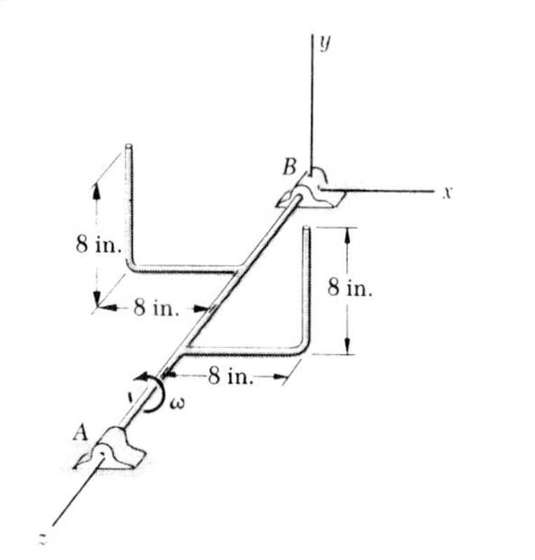

Fig. P18.C3

18.C3

Two L-shaped arms, each weighing 4 lb, are welded at the third points of a 2-ft horizontal shaft supported by bearings at A and B. The assembly is at rest ($\omega = 0$) when a couple of moment $\mathbf{M}_0 = (20 \text{ lb-in})\mathbf{k}$ is applied to the shaft. Determine the dynamic reactions exerted by the bearings on the axle at any time t after the couple has been applied. Resolve these reactions into components directed along the x and y axes rotating with the assembly. Plot the components of the reactions and the magnitudes of the reactions from
$t = 0$ to $t = 0.5$ s.

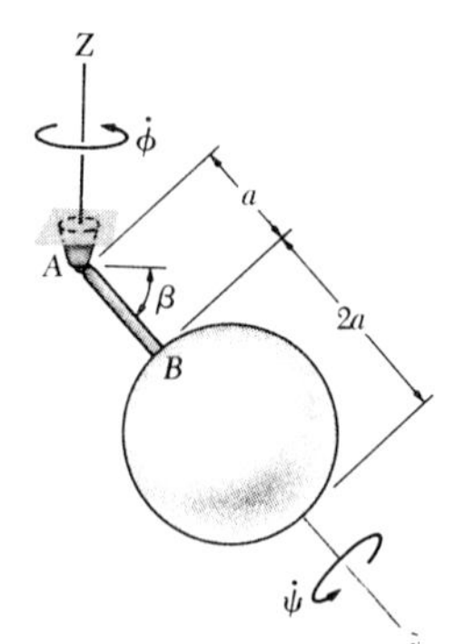

Fig. P18.C4

18.C4

A homogeneous sphere of mass m and radius a is welded to a rod AB of length a and negligible mass, which is held by a ball-and-socket support at A. Determine and plot the rates of steady precession $\dot{\phi}$ for values of β from zero to 90° and for values of spin rate $\dot{\psi}$ of $\sqrt{g/a}$, $5\sqrt{g/a}$, and $10\sqrt{g/a}$. Show that if the rate of spin is high, the positive steady precession rate is nearly equal to $5g/a\dot{\psi}$, independent of β.

18.C5

A homogeneous sphere of mass m and radius a is welded to a rod AB of length a and negligible mass, which is held by a ball-and-socket support at A. The sphere is released in the position $\beta = 0$ with a rate of precession $\dot{\phi}_0 = k\sqrt{g/a}$ and with no nutation. Determine and plot the largest value of β in the ensuing motion for values of k from 0.5 to 2 and for initial spin rates of $\dot{\psi}_0 =$ zero, $+\sqrt{g/a}$, and $-\sqrt{g/a}$.

CHAPTER 19
MECHANICAL VIBRATIONS

SECTIONS 19.1 to 19.4

19.1 A particle is known to move with a simple harmonic motion. The maximum acceleration is 3 m/s^2, and the maximum velocity is 150 mm/s. Determine the amplitude and the frequency of the motion.

19.2 Determine the maximum velocity and maximum acceleration of a particle which moves in simple harmonic motion with an amplitude of 150 mm and a period of 0.9 s.

19.3 A 5-lb collar is attached to a spring of constant 3 lb/in. and may slide without friction on a horizontal rod. If the collar is moved 4 in. from its equilibrium position and released, determine the maximum velocity and the maximum acceleration of the collar during the resulting motion.

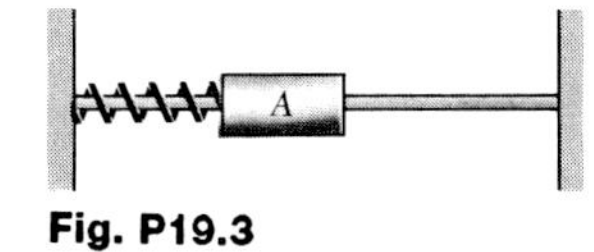

Fig. P19.3

19.4 A variable-speed motor is rigidly attached to the beam BC. The rotor is slightly unbalanced and causes the beam to vibrate with a frequency equal to the motor speed. When the speed of the motor is less than 450 rpm or more than 900 rpm, a small object placed at A is observed to remain in contact with the beam. For speeds between 450 and 900 rpm the object is observed to "dance" and actually to lose contact with the beam. Determine the amplitude of the motion of A when the speed of the motor is (*a*) 450 rpm, (*b*) 900 rpm. Give answers in both SI and U.S. customary units.

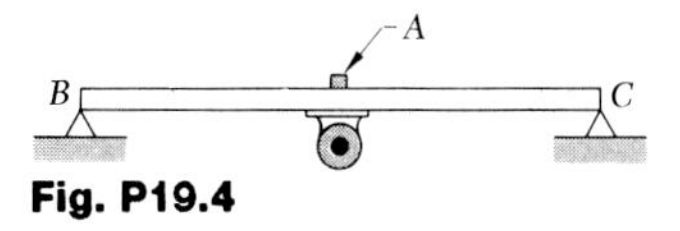

Fig. P19.4

19.5 An instrument package B is placed on the shaking table C as shown. The table is made to move horizontally in simple harmonic motion with a frequency of 3 Hz. Knowing that the coefficient of static friction is $\mu_s = 0.40$ between the package and the table, determine the largest allowable amplitude of the motion if the package is not to slip on the table. Give answers in both SI and U.S. customary units.

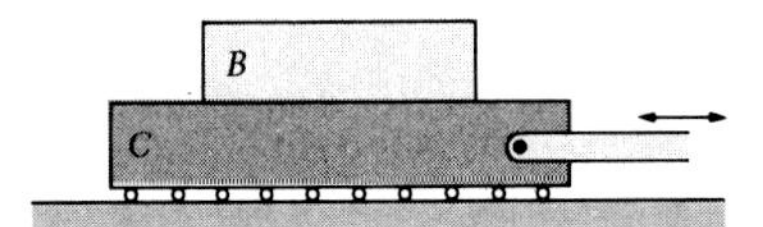

Fig. P19.5

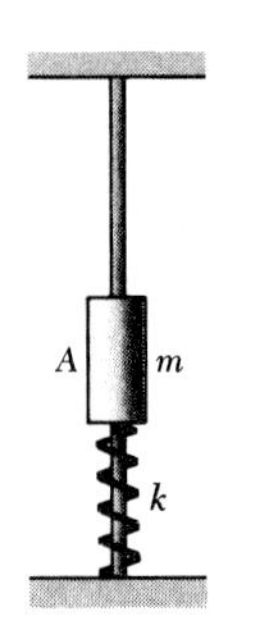

Fig. P19.6 and P19.7

19.6 A 4-kg collar is attached to a spring of constant $k = 800$ N/m as shown. If the collar is given a displacement of 40 mm downward from its equilibrium position and released, determine (*a*) the time required for the collar to move 60 mm upward, (*b*) the corresponding velocity and acceleration of the collar.

19.7 A 3-lb collar is attached to a spring of constant $k = 5$ lb/in. as shown. If the collar is given a displacement of 2.5 in. downward from its equilibrium position and released, determine (*a*) the time required for the collar to move 2 in. upward, (*b*) the corresponding velocity and acceleration of the collar.

19.8 In Prob. 19.7, determine the position, velocity, and acceleration of the collar 0.20 s after it has been released.

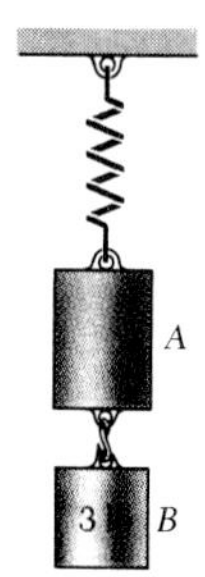

Fig. P19.9 and P19.10

19.9 The period of vibration of the system shown is observed to be 0.4 s. After cylinder B has been removed, the period is observed to be 0.3 s. Determine (*a*) the weight of cylinder A, (*b*) the constant of the spring.

19.10 The period of vibration of the system shown is observed to be 1.5 s. After cylinder B has been removed and replaced with a 4-lb cylinder, the period is observed to be 1.6 s. Determine (*a*) the weight of cylinder A, (*b*) the constant of the spring.

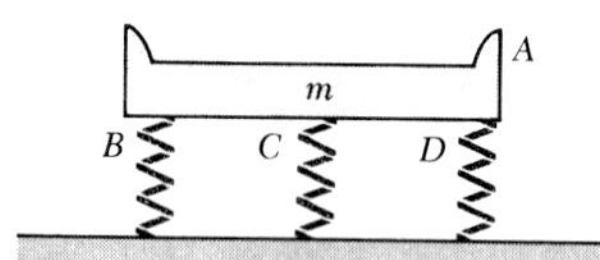

Fig. P19.11 and P19.12

19.11 A tray of mass m is attached to three springs as shown. The period of vibration of the empty tray is 0.5 s. After a 1.5-kg block has been placed in the center of the tray, the period is observed to be 0.6 s. Knowing that the amplitude of the vibration is small, determine the mass m of the tray.

19.12 A tray of mass m is attached to three springs as shown. The period of vibration of the empty tray is 0.75 s. After the center spring C has been removed, the period is observed to be 0.90 s. Knowing that the constant of spring C is 100 N/m, determine the mass m of the tray.

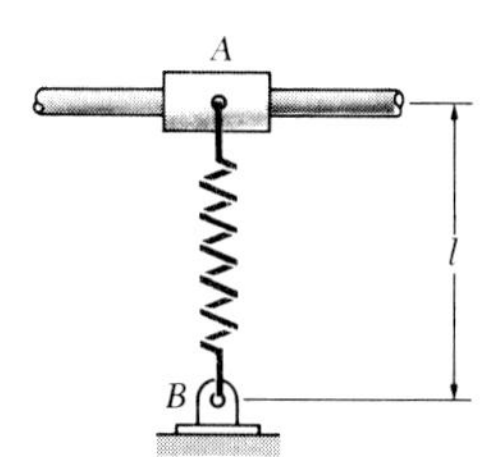

Fig. P19.13

19.13 A collar of mass m slides without friction on a horizontal rod and is attached to a spring AB of constant k. (*a*) If the unstretched length of the spring is just equal to l, show that the collar does not execute simple harmonic motion even when the amplitude of the oscillations is small. (*b*) If the unstretched length of the spring is less than l, show that the motion may be approximated by a simple harmonic motion for small oscillations.

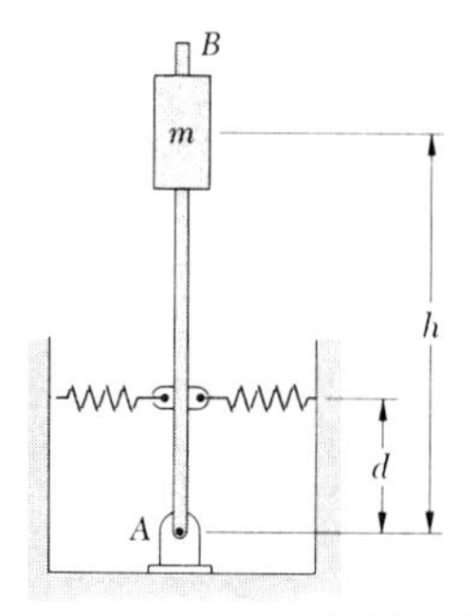

Fig. P19.14 and P19.15

19.14 The rod AB is attached to a hinge at A and to two springs each of constant k. If $h = 600$ mm, $d = 250$ mm, and $m = 25$ kg, determine the value of k for which the period of small oscillations is (*a*) 1 s, (*b*) infinite. Neglect the mass of the rod and assume that each spring can act in either tension or compression.

19.15 If $h = 600$ mm and $d = 400$ mm and each spring has a constant $k = 700$ N/m, determine the mass m for which the period of small oscillations is (*a*) 0.50 s, (*b*) infinite. Neglect the mass of the rod and assume that each spring can act in either tension or compression.

19.16 Using a table of elliptic integrals, determine the period of a simple pendulum of length $l = 750$ mm if the amplitude of the oscillations is $\theta_m = 50°$.

19.17 The 3-kg uniform rod shown is attached to a spring of constant $k = 900$ N/m. If end A of the rod is depressed 25 mm and released, determine (a) the period of vibration, (b) the maximum velocity of end A.

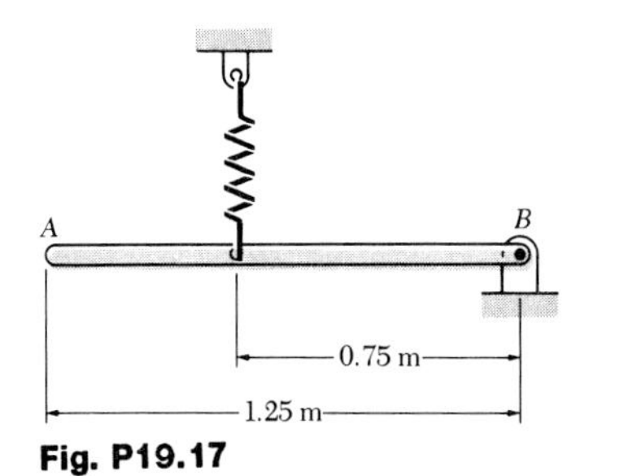

Fig. P19.17

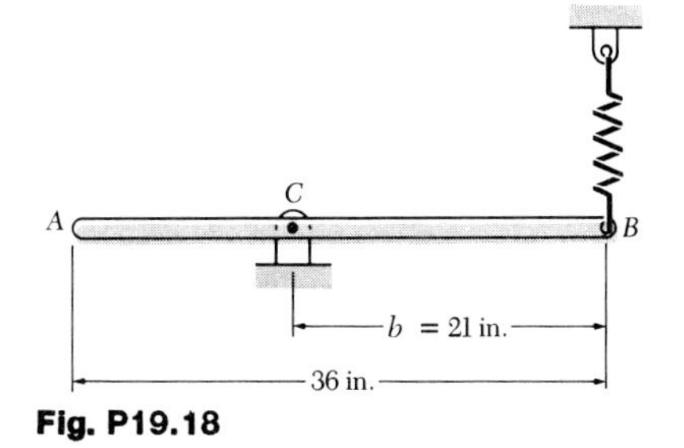

Fig. P19.18

19.18 The uniform rod shown weighs 12 lb and is attached to a spring of constant $k = 3$ lb/in. If end B of the rod is depressed 0.5 in. and released, determine (a) the period of vibration, (b) the maximum velocity of end B.

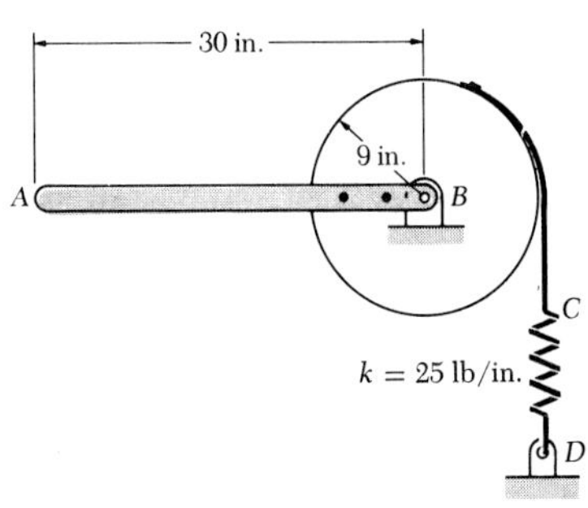

Fig. P19.19

19.19 A 12-lb slender rod AB is riveted to a 10-lb uniform disk as shown. A belt is attached to the rim of the disk and to a spring which holds the rod at rest in the position shown. If end A of the rod is moved 1.5 in. down and released, determine (a) the period of vibration, (b) the maximum velocity of end A.

19.20 A belt is placed over the rim of a 12-kg disk as shown and then attached to a 4-kg cylinder and to a spring of constant $k = 500$ N/m. If the cylinder is moved 75 mm down from its equilibrium position and released, determine (a) the period of vibration, (b) the maximum velocity of the cylinder. Assume friction is sufficient to prevent the belt from slipping on the rim.

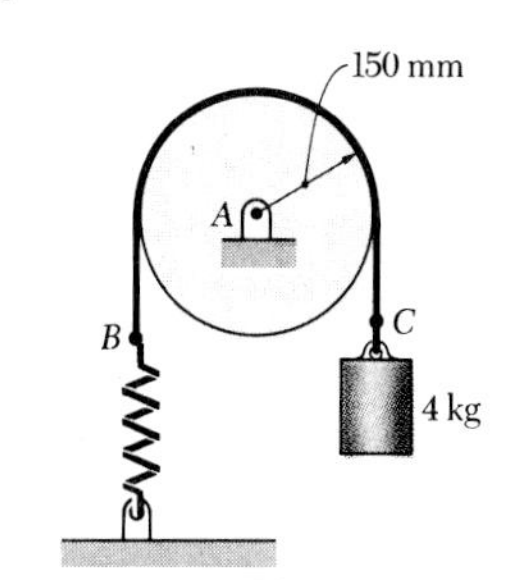

Fig. P19.20

19.21 In Prob. 19.20, determine (a) the frequency of vibration, (b) the maximum tension which occurs in the belt at B and at C.

19.22 In Prob. 19.18, determine (a) the value of b for which the smallest period of vibration occurs, (b) the corresponding period of vibration.

19.23 The 3-kg uniform rod AB is attached as shown to a spring of constant $k = 900$ N/m. A small 0.5-kg block C is placed on the rod at A. (a) If end A of the rod is then moved down through a small distance δ_0 and released, determine the period of the vibration. (b) Determine the largest allowable value of δ_0 if block C is to remain at all times in contact with the rod.

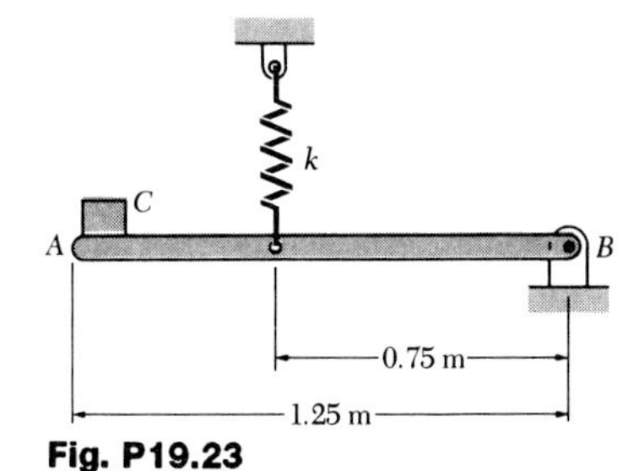

Fig. P19.23

19.24 A slender rod of mass m and length l is held by two springs, each of constant k. Determine the frequency of the resulting vibration if the rod is (a) given a small vertical displacement and released, (b) rotated through a small angle about a horizontal axis through G and released. (c) Determine the ratio b/l for which the frequencies found in parts a and b are equal.

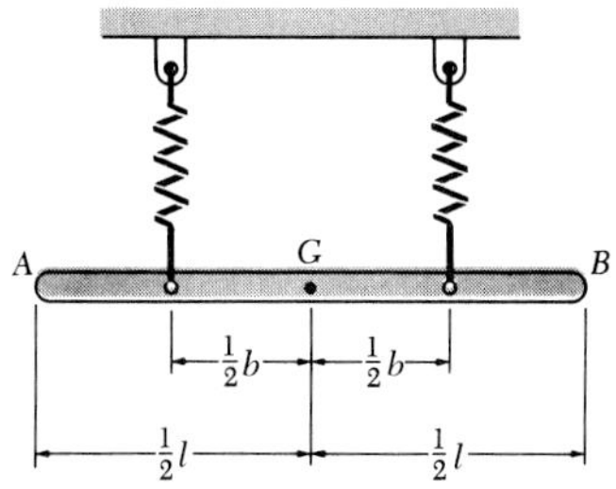

Fig. P19.24

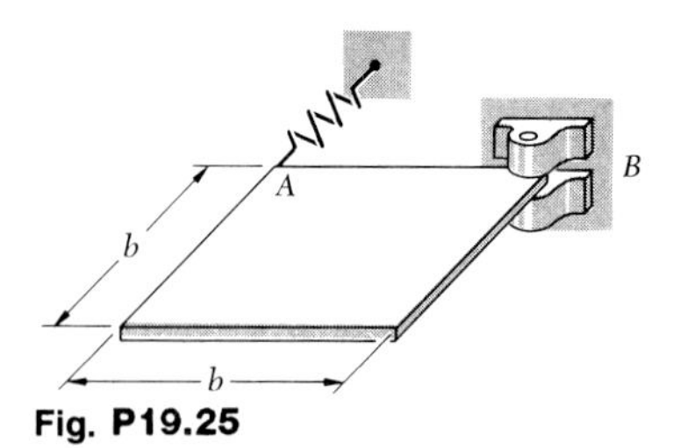

Fig. P19.25

19.25 A uniform square plate of mass m is supported in a horizontal plane by a vertical pin at B and is attached at A to a spring of constant k. If corner A is given a small displacement and released, determine the period of the resulting motion.

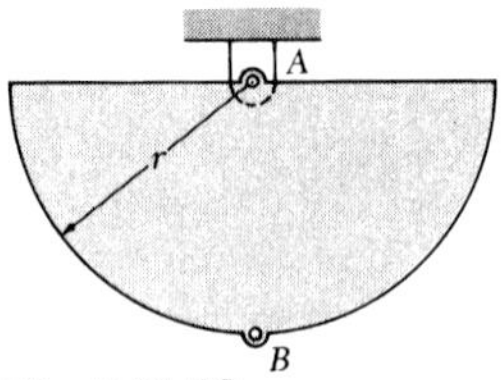

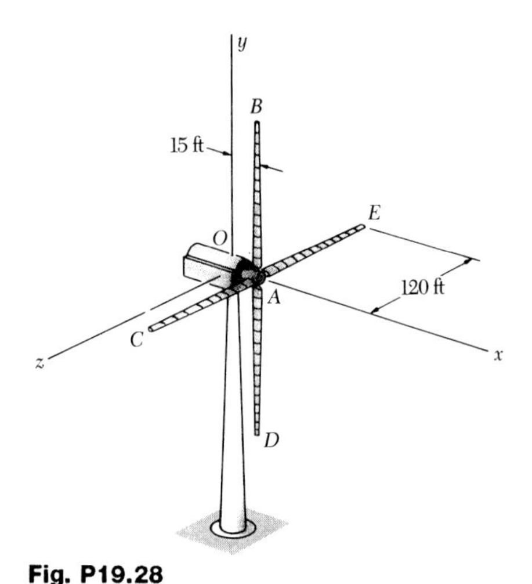

Fig. P19.26

19.26 A uniform semicircular disk of radius r is suspended from a hinge. Determine the period of small oscillations if the disk (a) is suspended from A as shown, (b) is suspended from point B.

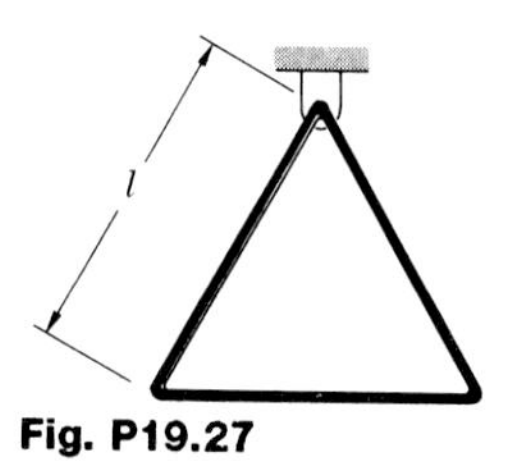

Fig. P19.27

19.27 A thin homogeneous wire is bent into the shape of an equilateral triangle of side $l = 250$ mm. Determine the period of small oscillations if the wire (a) is suspended as shown, (b) is suspended from a pin located at the midpoint of one side.

Fig. P19.28

19.28 Blade AB of the wind-turbine generator shown is to be temporarily removed. Motion of the turbine generator about the y axis is prevented, but the remaining three blades may oscillate as a unit about the x axis. Assuming that each blade is equivalent to a 120-ft slender rod, determine the period of small oscillations of the blades in the absence of wind.

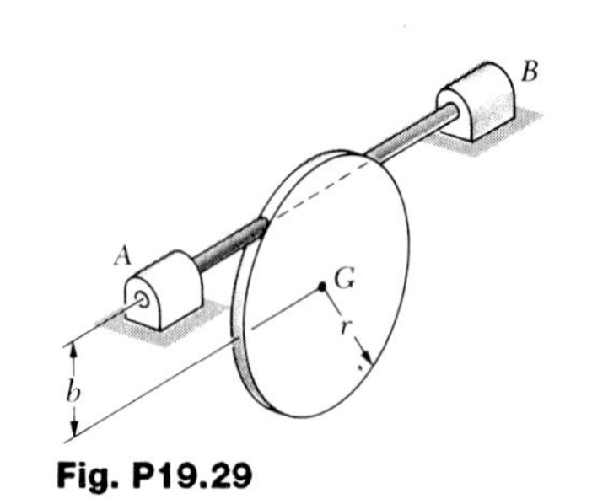

Fig. P19.29

19.29 A thin disk of radius r may oscillate about the axis AB located as shown at a distance b from the mass center G. (a) Determine the period of small oscillations if $b = r$. (b) Determine a second value of b for which the period of oscillation is the same as that found in part a.

19.30 A 6-kg slender rod is suspended from a steel wire which is known to have a torsional spring constant $K = 1.75$ N·m/rad. If the rod is rotated through 180° about the vertical and then released, determine (a) the period of oscillation, (b) the maximum velocity of end A of the rod.

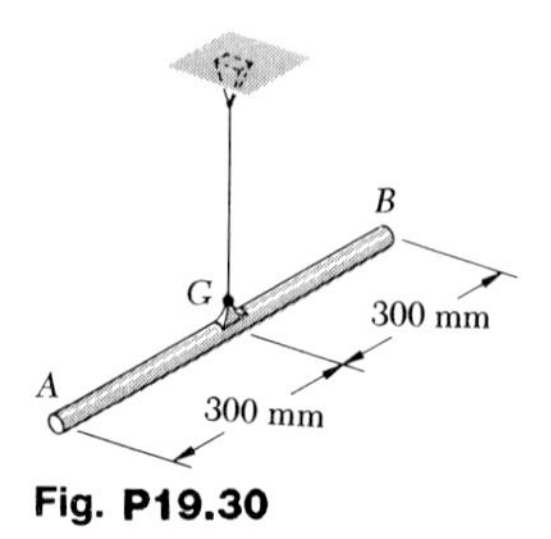

Fig. P19.30

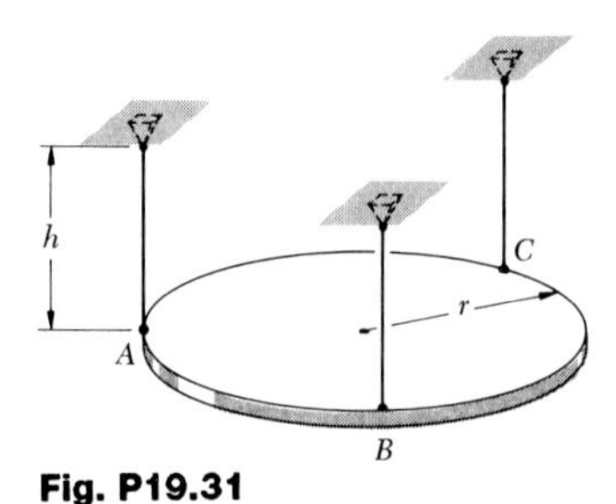

Fig. P19.31

19.31 A thin circular plate of radius r is suspended from three vertical wires of length h equally spaced around the perimeter of the plate. Determine the period of oscillation when (a) the plate is rotated through a small angle about a vertical axis passing through its mass center and released, (b) the plate is given a small horizontal translation and released.

19.32 A slender bar of length l is attached by a smooth pin A to a collar of negligible mass. Determine the period of small oscillations of the bar, assuming that the coefficient of friction between the collar and the horizontal rod (a) is sufficient to prevent any movement of the collar, (b) is zero.

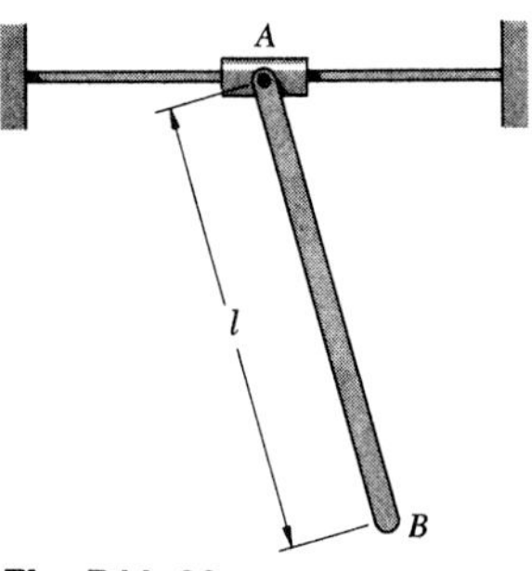

Fig. P19.32

SECTION 19.6

19.33 A homogeneous wire of length $2l$ is bent as shown and allowed to oscillate about a frictionless pin at B. Denoting by τ_0 the period of small oscillations when $\beta = 0$, determine the angle β for which the period of small oscillations is $2\tau_0$.

19.34 Knowing that $l = 750$ mm and $\beta = 40°$, determine the period of small oscillations of the bent homogeneous wire shown.

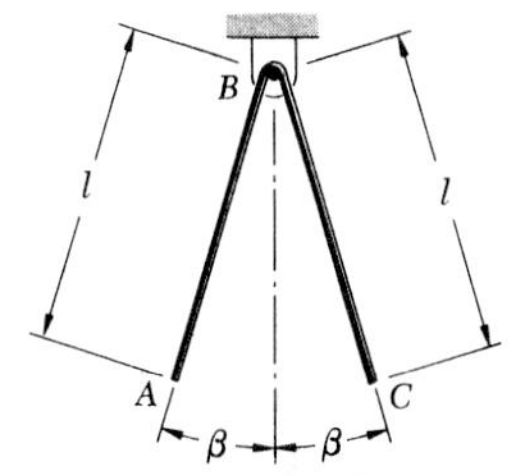

Fig. P19.33 and P19.34

19.35 A uniform disk of radius c is supported by a ball-and-socket joint at A. Determine the frequency of small oscillations of the disk (a) in the plane of the disk, (b) in a direction perpendicular to the disk.

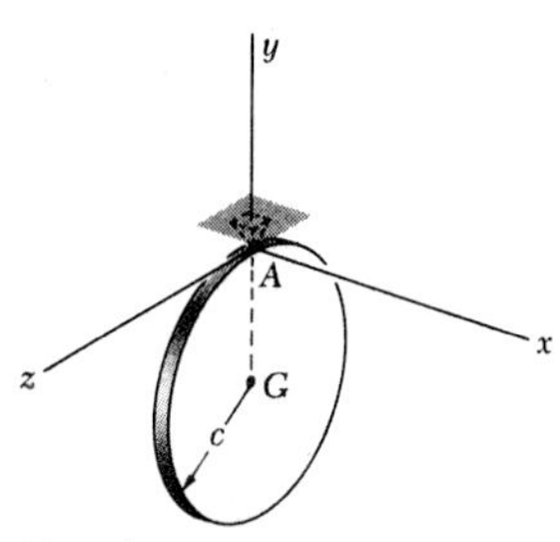

Fig. P19.35

19.36 It is observed that when an 8-lb weight is attached to the rim of a 6-ft-diameter flywheel, the period of small oscillations of the flywheel is 22 s. Neglecting axle friction, determine the centroidal moment of inertia of the flywheel.

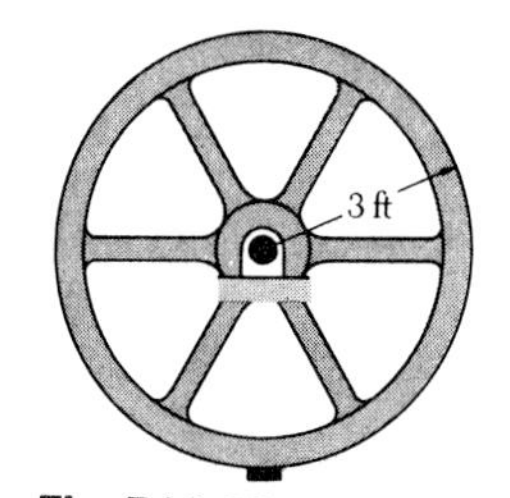

Fig. P19.36

19.37 The uniform rod ABC weighs 5 lb and is attached to two springs as shown. If end C is given a small displacement and released, determine the frequency of vibration of the rod.

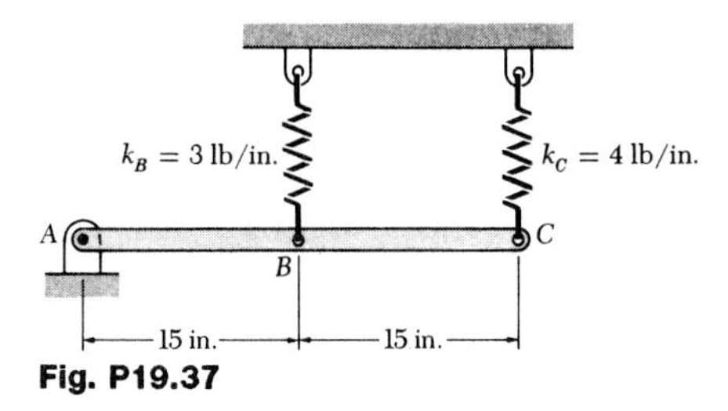

Fig. P19.37

19.38 For the rod of Prob. 19.37, determine the frequency of vibration of the rod if the springs are interchanged so that $k_B = 4$ lb/in. and $k_C = 3$ lb/in.

19.39 The 5-kg slender rod AB is welded to the 8-kg uniform disk. A spring of constant 450 N/m is attached to the disk and holds the rod at rest in the position shown. If end B of the rod is given a small displacement and released, determine the period of vibration of the rod.

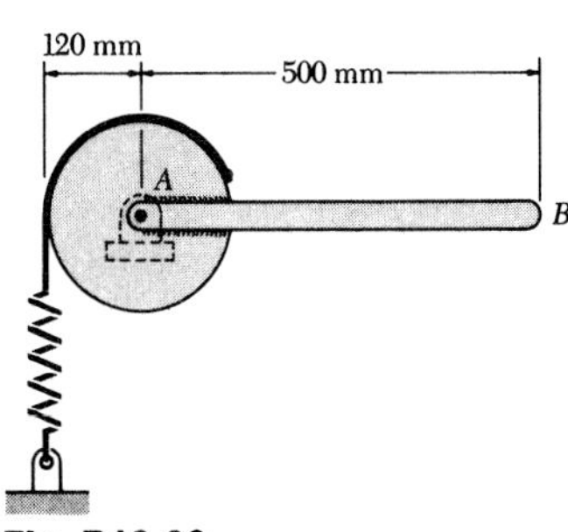

Fig. P19.39

19.40 For the rod and disk of Prob. 19.39, determine the constant of the spring for which the period of vibration of the rod is 1.5 s.

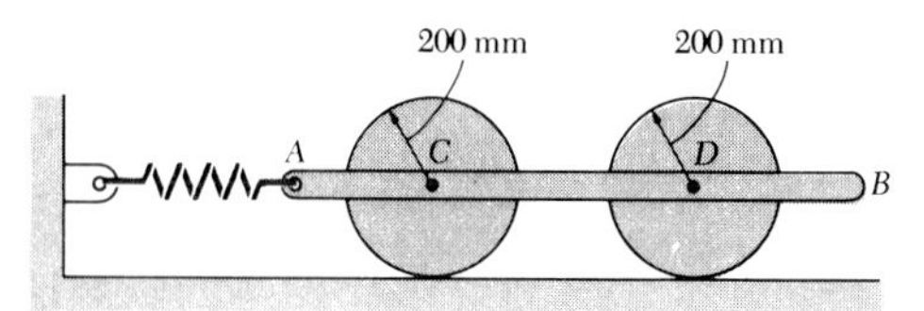

Fig. P19.41

19.41 Two 5-kg uniform disks are attached to the 8-kg rod AB as shown. Knowing that the constant of the spring is 4 kN/m and that the disks roll without sliding, determine the frequency of vibration of the system.

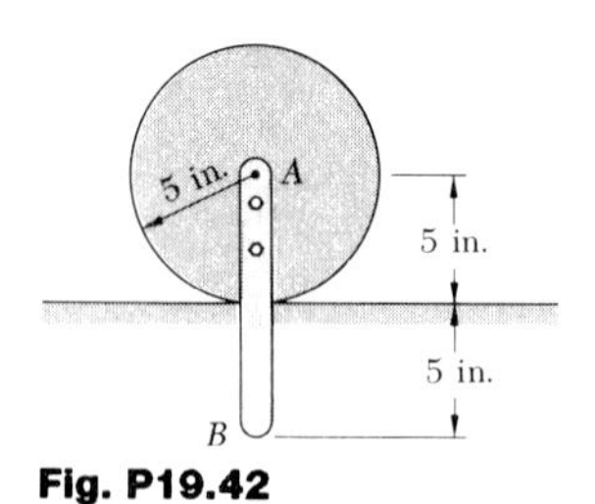

Fig. P19.42

19.42 The 4-lb rod AB is bolted to the 5-in.-radius disk as shown. Knowing that the disk rolls without sliding, determine the weight of the disk for which the period of small oscillations of the system is 1.5 s.

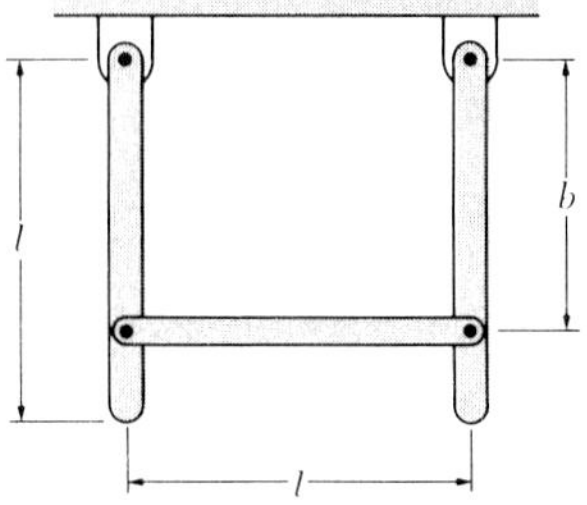

Fig. P19.43 and P19.44

19.43 Three identical rods are connected as shown. If $b = \frac{3}{4}l$, determine the frequency of small oscillations of the system.

19.44 Three identical rods are connected as shown. Determine (*a*) the distance b for which the frequency of oscillation is maximum, (*b*) the corresponding maximum frequency.

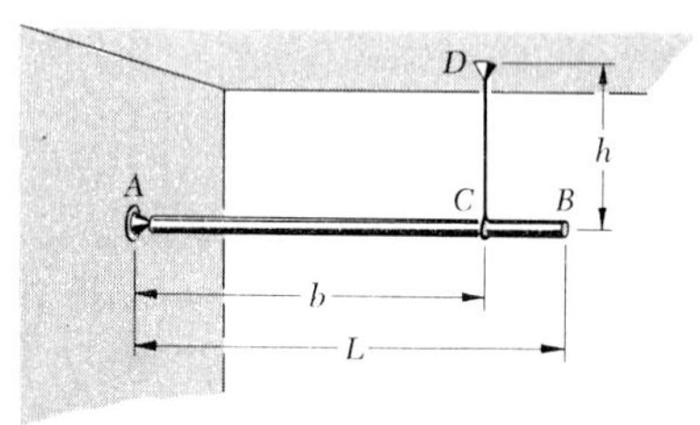

Fig. P19.45

19.45 A uniform rod of length L is supported by a ball-and-socket joint at A and by the vertical wire CD. Derive an expression for the period of oscillation of the rod if end B is given a small horizontal displacement and then released.

19.46 Solve Prob. 19.45, assuming that $L = 3$ m, $b = 2.5$ m, and $h = 2$ m.

19.47 A half section of pipe is placed on a horizontal surface, rotated through a small angle, and then released. Assuming that the pipe section rolls without sliding, determine the period of oscillation.

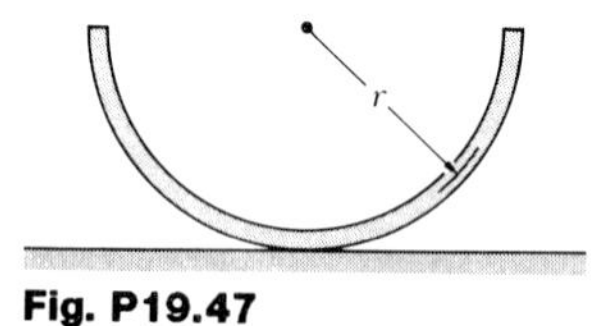

Fig. P19.47

19.48 The 10-kg rod AB is attached to 4-kg disks as shown. Knowing that the disks roll without sliding, determine the frequency of small oscillations of the system.

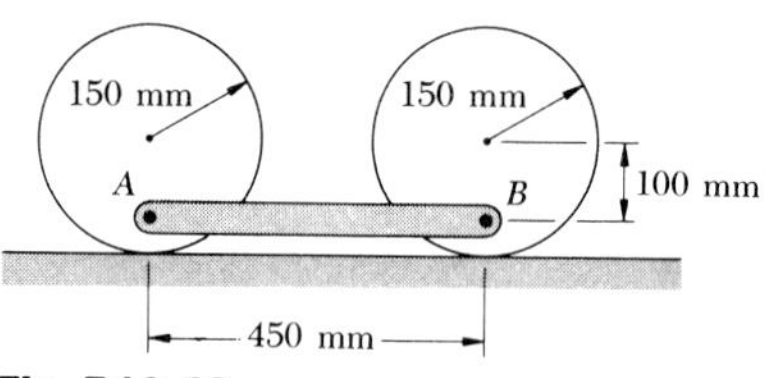

Fig. P19.48

SECTION 19.7

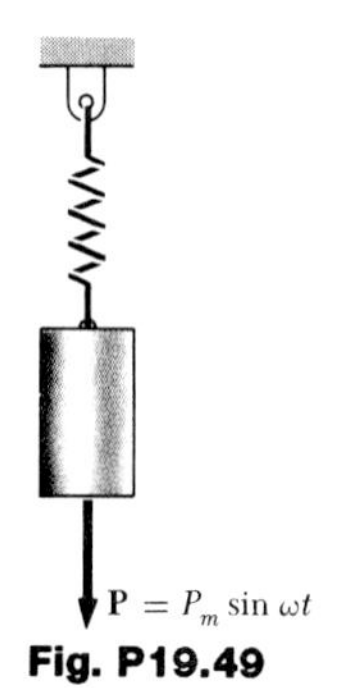

Fig. P19.49

19.49 A 5-kg cylinder is suspended from a spring of constant 320 N/m and is acted upon by a vertical periodic force of magnitude $P = P_m \sin \omega t$, where $P_m = 14$ N. Determine the amplitude of the motion of the cylinder if (*a*) $\omega = 6$ rad/s, (*b*) $\omega = 12$ rad/s.

19.50 A cylinder of mass m is suspended from a spring of constant k and is acted upon by a vertical periodic force of magnitude $P = P_m \sin \omega t$. Determine the range of values of ω for which the amplitude of the vibration exceeds twice the static deflection caused by a constant force of magnitude P_m.

19.51 In Prob. 19.50, determine the range of values of ω for which the amplitude of the vibration is less than the static deflection caused by a constant force of magnitude P_m.

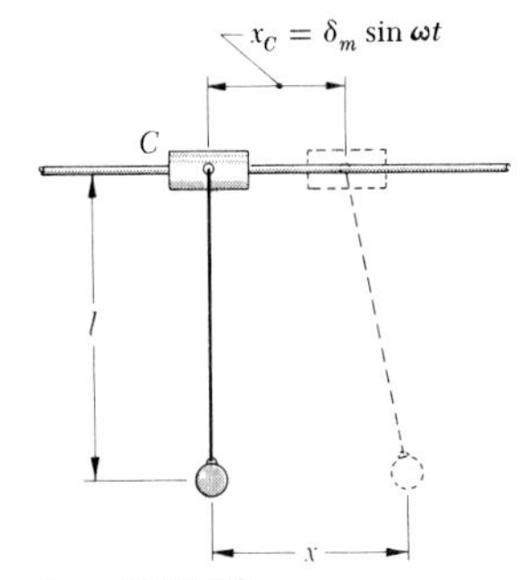

Fig. P19.52

19.52 A simple pendulum of length l is suspended from a collar C which is forced to move horizontally according to the relation $x_C = \delta_m \sin \omega t$. Determine the range of values of ω for which the amplitude of the motion of the bob is less than δ_m. (Assume that δ_m is small compared with the length l of the pendulum.)

19.53 In Prob. 19.52, determine the range of values of ω for which the amplitude of the motion of the bob exceeds $3\delta_m$.

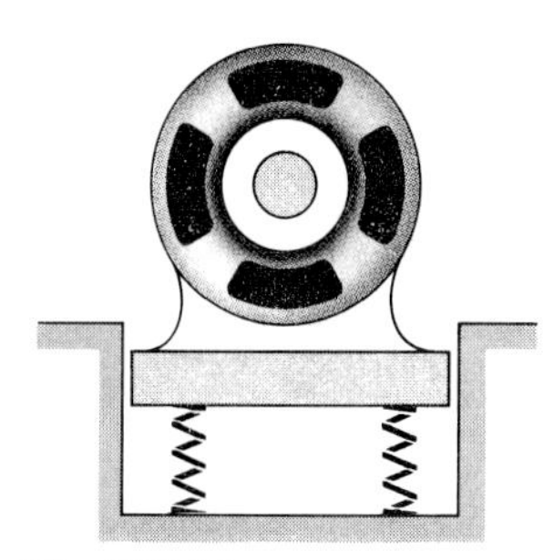
Fig. P19.56

19.54 As the speed of a spring-supported motor is slowly increased from 300 to 400 rpm, the amplitude of the vibration due to the unbalance of the rotor is observed to decrease continuously from 0.075 to 0.040 in. Determine the speed at which resonance will occur.

19.55 For the spring-supported motor of Prob. 19.54, determine the speed of the motor for which the amplitude of the vibration is 0.100 in.

19.56 A motor of mass 9 kg is supported by four springs, each of constant 20 kN/m. The motor is constrained to move vertically, and the amplitude of its motion is observed to be 1.2 mm at a speed of 1200 rpm. Knowing that the mass of the rotor is 2.5 kg, determine the distance between the mass center of the rotor and the axis of the shaft.

19.57 In Prob. 19.56, determine the amplitude of the vertical motion of the motor at a speed of (*a*) 450 rpm, (*b*) 1600 rpm, (*c*) 900 rpm.

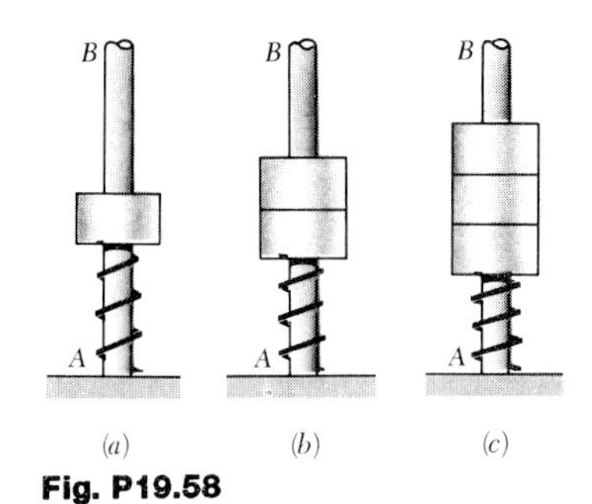

Fig. P19.58

19.58 Rod AB is rigidly attached to the frame of a motor running at a constant speed. When a collar of mass m is placed on the spring, it is observed to vibrate with an amplitude of 10 mm. When two collars, each of mass m, are placed on the spring, the amplitude is observed to be 12 mm. What amplitude of vibration should be expected when three collars, each of mass m, are placed on the spring? (Obtain two answers.)

19.59 Solve Prob. 19.58, assuming that the speed of the motor is changed and that one collar has an amplitude of 12 mm and two collars have an amplitude of 4 mm.

19.60 Three identical cylinders A, B, and C are suspended from a bar DE by two or more identical springs as shown. Bar DE is known to move vertically according to the relation $y = \delta_m \sin \omega t$. Knowing that the amplitudes of vibration of cylinders A and B are, respectively, 1.5 in. and 0.75 in., determine the expected amplitude of vibration of cylinder C. (Obtain two answers.)

19.61 Solve Prob. 19.60, assuming that the amplitudes of vibration of cylinders A and B are, respectively, 0.8 in. and 1.2 in.

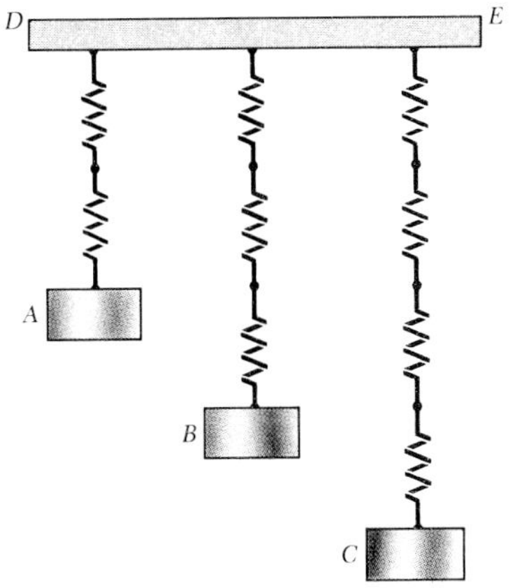

Fig. P19.60

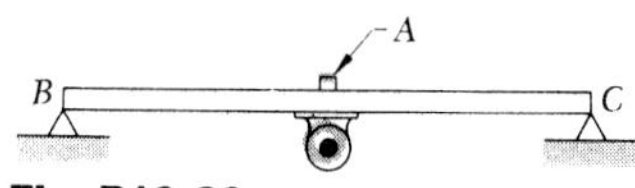

Fig. P19.62

19.62 A variable-speed motor is rigidly attached to the beam BC. When the speed of the motor is less than 750 rpm or more than 1500 rpm, a small object placed at A is observed to remain in contact with the beam. For speeds between 750 and 1500 rpm the object is observed to "dance" and actually to lose contact with the beam. Determine the speed at which resonance will occur.

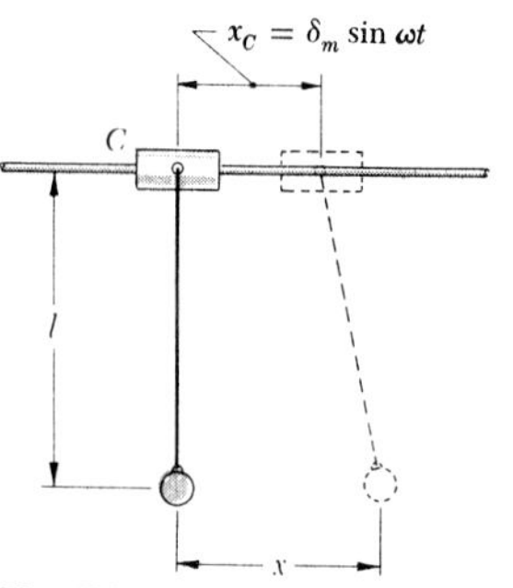

Fig. P19.63 and P19.64

19.63 The amplitude of the motion of the pendulum bob shown is observed to be 60 mm, when the amplitude of the motion of collar C is 15 mm. Knowing that the length of the pendulum is $l = 0.90$ m, determine the two possible values of the frequency of the horizontal motion of collar C.

19.64 A pendulum is suspended from collar C as shown. Knowing that the amplitude of the motion of the bob is 12 mm for $l = 750$ mm and 17 mm for $l = 500$ mm, determine the frequency and amplitude of the horizontal motion of collar C.

SECTION 19.8

19.65 The barrel of a field gun weighs 1400 lb and is returned into firing position after recoil by a recuperator of constant $k = 10{,}000$ lb/ft. Determine (*a*) the value of the coefficient of damping of the recuperator which causes the barrel to return into firing position in the shortest possible time without oscillation, (*b*) the time needed for the barrel to move from its maximum-recoil position halfway back to its firing position.

19.66 Assuming that the barrel of the gun of Prob. 19.65 is modified, with a resulting increase in weight of 400 lb, determine (*a*) the constant k which should be used for the recuperator if the barrel is to remain critically damped, (*b*) the time needed for the modified barrel to move from its maximum-recoil position halfway back to its firing position.

19.67 In the case of the forced vibration of a system with a given damping factor c/c_c, determine the frequency ratio ω/p for which the amplitude of the vibration is maximum.

19.68 A 30-lb motor is directly supported by a light horizontal beam which has a static deflection of 0.05 in. due to the weight of the motor. Knowing that the unbalance of the rotor is equivalent to a weight of 1 oz located 7.5 in. from the axis of rotation, determine the amplitude of vibration of the motor at a speed of 900 rpm, assuming (*a*) that no damping is present, (*b*) that the damping factor c/c_c is equal to 0.075.

19.69 A motor weighing 50 lb is supported by four springs, each having a constant of 1000 lb/in. The unbalance of the rotor is equivalent to a weight of 1 oz located 5 in. from the axis of rotation. Knowing that the motor is constrained to move vertically, determine the amplitude of the steady-state vibration of the motor at a speed of 1800 rpm, assuming (*a*) that no damping is present, (*b*) that the damping factor c/c_c is equal to 0.125.

19.70 Solve Prob. 19.56, assuming that a dashpot having a coefficient of damping $c = 200$ N·s/m has been connected to the motor and to the ground.

19.71 A 50-kg motor is directly supported by a light horizontal beam which has a static deflection of 6 mm due to the weight of the motor. The unbalance of the rotor is equivalent to a mass of 100 g located 75 mm from the axis of rotation. Knowing that the amplitude of the vibration of the motor is 0.9 mm at a speed of 400 rpm, determine (*a*) the damping factor c/c_c, (*b*) the coefficient of damping c.

19.72 A machine element having a mass of 400 kg is supported by two springs, each having a constant of 38 kN/m. A periodic force of maximum magnitude equal to 135 N is applied to the element with a frequency of 2.5 Hz. Knowing that the coefficient of damping is 1400 N·s/m, determine the amplitude of the steady-state vibration of the element.

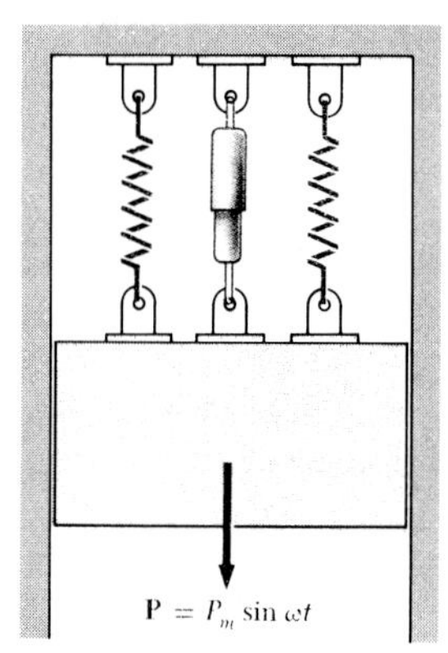

Fig. P19.72

19.73 In Prob. 19.72, determine the required value of the coefficient of damping if the amplitude of the steady-state vibration of the element is to be 3.5 mm.

19.74 A platform of weight 200 lb, supported by two springs each of constant $k = 250$ lb/in., is subjected to a periodic force of maximum magnitude equal to 125 lb. Knowing that the coefficient of damping is 10 lb·s/in., determine (*a*) the natural frequency in rpm of the platform if there were no damping, (*b*) the frequency in rpm of the periodic force corresponding to the maximum value of the magnification factor, assuming damping, (*c*) the amplitude of the actual motion of the platform for each of the frequencies found in parts *a* and *b*.

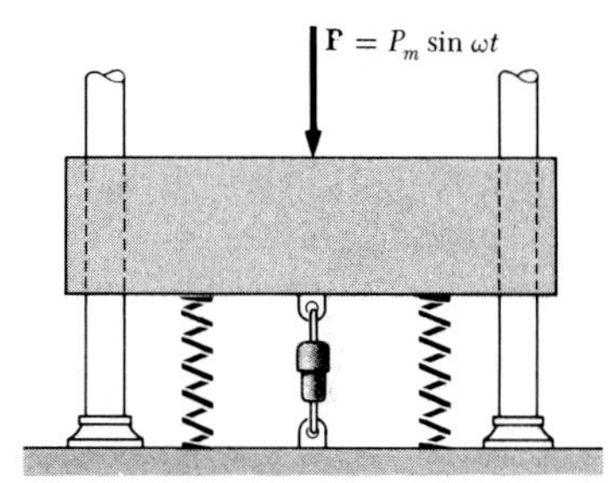

Fig. P19.74

19.75 Solve Prob. 19.74, assuming that the coefficient of damping is increased to 15 lb·s/in.

19.76 Two blocks *A* and *B*, each of mass *m*, are suspended as shown by means of five springs of the same constant *k* and are connected by a dashpot of coefficient of damping *c*. Block *B* is subjected to a force of magnitude $P = P_m \sin \omega t$. Write the differential equations defining the displacements x_A and x_B of the two blocks from their equilibrium positions.

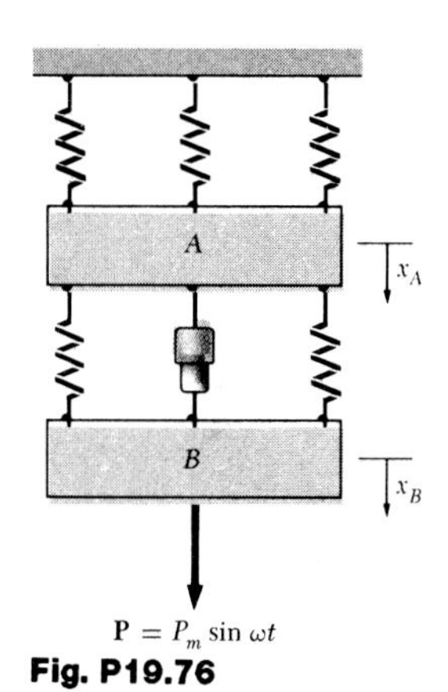

Fig. P19.76

19.77 Draw the electrical analogue of the mechanical system shown. (*Hint.* Draw the loops corresponding to the free bodies *m* and *A*.)

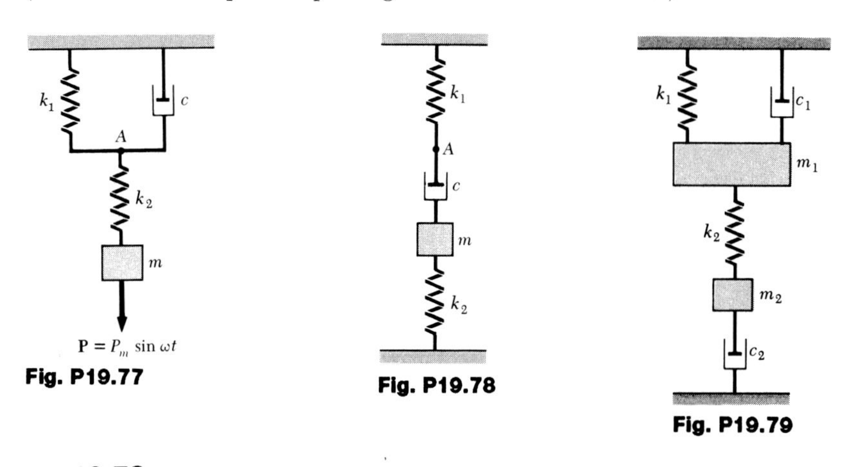

Fig. P19.77

Fig. P19.78

Fig. P19.79

19.78 Write the differential equations defining (*a*) the displacements of mass *m* and point *A*, (*b*) the currents in the corresponding loops of the electrical analogue.

19.79 Draw the electrical analogue of the mechanical system shown.

19.80 Write the differential equations defining (*a*) the displacements of the masses m_1 and m_2, (*b*) the currents in the corresponding loops of the electrical analogue.

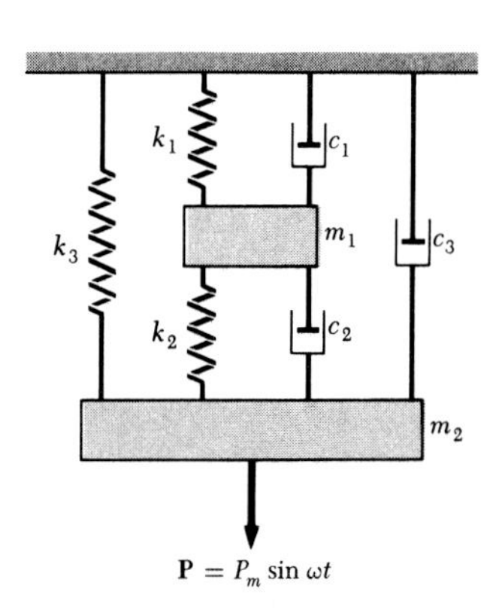

Fig. P19.80

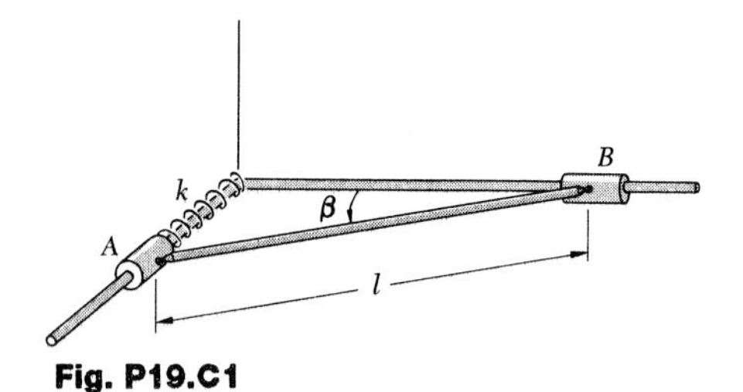

Fig. P19.C1

19.C1

The slender rod AB of mass m is attached to two identical collars, each of mass M. The system lies in a horizontal plane and is in equilibrium in the position shown when the collar A is given a small displacement and released. Determine and plot the period of vibration as a function of the angle β from 10° to 80° for mass ratios M/m of 0, 1, and 5.

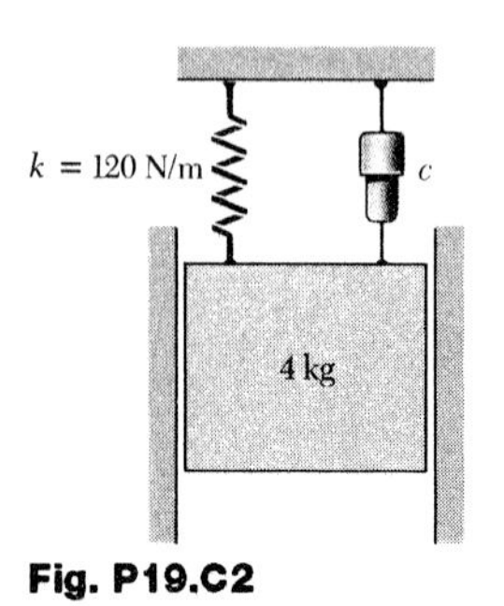

Fig. P19.C2

19.C2

The block shown is depressed 20 mm from its equilibrium position and released from rest at $t = 0$. Determine and plot the displacement as a function of time from $t = 0$ to $t = 2$ seconds for damping factors $c / c_c = 0.4$, 1.0, and 1.6.

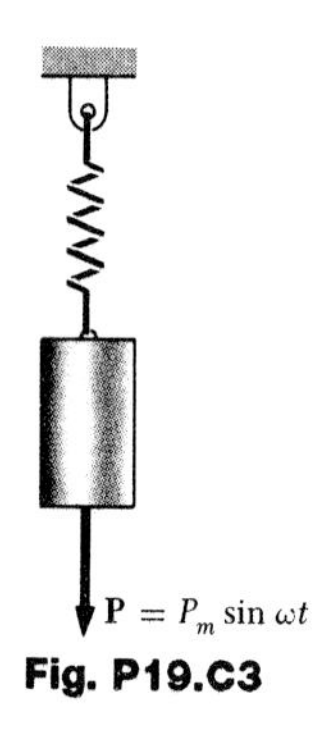

Fig. P19.C3

19.C3

A 5-kg cylinder is suspended from a spring of constant 320 N/m and is at rest in the static equilibrium position at time $t = 0$. The cylinder is acted upon by a vertical periodic force of magnitude $P_m \sin \omega t$ where $P_m = 14$ N. Write the differential equation of motion and obtain the general solution. Plot the displacement as a function of time from $t = 0$ to $t = 8$ seconds if (a) $\omega = 6$ rad/s and (b) $\omega = 10$ rad/s. Show that in both cases the displacement is a periodic function of frequency 2 rad/s, the "beat" frequency.

19.C4

The suspension of an automobile may be approximated by the simplified spring-and-dashpot system shown. (a) Write the differential equation defining the motion of the mass m relative to the road when the system moves at a speed v over a road of sinusoidal cross section as shown. (b) Derive an expression for the amplitude of the relative motion of the mass. (c) Plot the amplitude of the relative motion as a function of the speed v for damping factors $c / c_c = 0.125$, 0.250, 0.5, and 1.

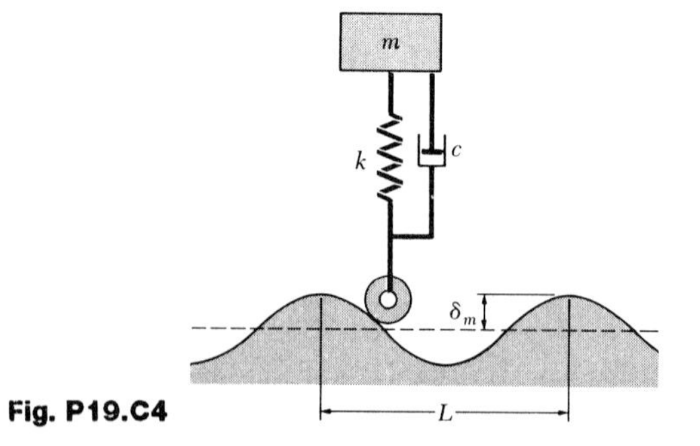

Fig. P19.C4

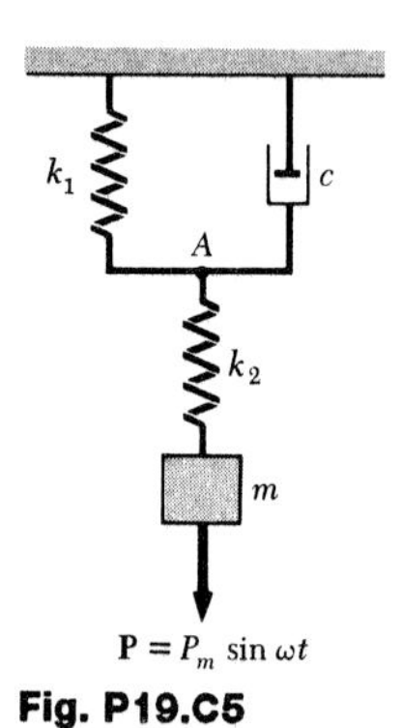

Fig. P19.C5

19.C5

A block of weight 200 lb, supported by two springs of constants $k_1 = k_2 = 500$ lb/in., is subjected to a periodic force of maximum magnitude $P_m = 125$ lb. The coefficient of damping is $c = 15$ lb-s/in. Determine and plot the amplitude of the steady state vibration of the block for values of ω from zero to 100 rad/s. Determine the maximum amplitude and the corresponding frequency.

ANSWERS TO ODD-NUMBERED PROBLEMS

CHAPTER 11

11.1 $x = 22$ m; $v = 12$ m/s; $a = 18$ m/s^2.
11.3 $t = 1$s; $x = 25$ m; $a = -12$ m/s^2; or $t = 5$ s; $x = -7$ m; $a = 12$ m/s^2.
11.5 (*a*) 4 s. (*b*) $x = -22$ m; $v = -240$ m/s. (*c*) 328 m.
11.7 (*a*) 22.5 m^3/s^2. (*b*) 6.32 m/s.
11.9 12 s^{-2}.
11.11 (*a*) 20 in. (*b*) ∞. (*c*) 1.535 s.
11.15 (*a*) -2.43x10^6 ft/s^2. (*b*) 1.366x10^{-3} s.
11.17 (*a*) 6.38 m/s up. (*b*) 23.0 m/s down.
11.19 (*a*) 15 ft/s. (*b*) 65 ft/s. (*c*) 275 ft.
11.21 (*a*) 7.76 s. (*b*) 51.5 km/h.
11.23 (*b*) 1.529 s; 11.47 m.
11.25 (*a*) $t = 16$ s; $x = 960$ ft. (*b*) $v_A = 51.8$ mi/h; $v_B = 28.6$ mi/h.
11.27 (*a*) 30 in./s right. (*b*) 40 in./s right. (*c*) 10 in./s right. (*d*) 20 in./s left.
11.29 (*a*) $a_A = 50$ mm/s^2 left; $a_B = 25$ mm/s^2 Right.
(*b*) $v_B = 125$ mm/s left; $x_B = 312$ mm right.
11.31 (*a*) 2 s. (*b*) 1.5 in. down.
11.33 (*a*) 25 ft/s. (*b*) 114 ft.
11.35 29 s.
11.37 15 s.
11.39 (*a*) $t = 15$ s; $x = 225$ m. (*b*) $t = 21$ s; $x = 315$ m.
11.41 91.8 km/h.
11.43 (*a*) $v = 50$ m/s; $x = 996$ m. (*b*) 49.5 m/s.
11.45 (*a*) 8 s. (*b*) 3 ft/s.
11.47 (*a*) $v = 20.2$ mi/h; $x = 74.8$ ft. (*b*) $v = 25.5$ mi/h; $x = 176.1$ ft.
11.49 (*a*) -7.56 in./s^2. (*b*) -8.8 in./s^2.
11.51 (*a*) $\mathbf{v} = 2.83$ m/s at $-45°$. (*b*) $t = 1.5$ s; $x = y = -1.5$ m.
11.53 (*a*) 13.05 m/s. (*b*) 27.6 m.
11.55 4.70 m/s $\leq v_0 \leq$ 7.23 m/s.
11.57 $d_B = 8.55$ m; $d_C = 11.52$ m.
11.59 $\alpha = 32.3°$.
11.61 54.0 km/h at 31.3°.
11.63 (*a*) 91.0 ft/s at 243°. (*b*) 273 ft at 243°. (*c*) 434 ft.
11.65 (*a*) 63.9°. (*b*) $\mathbf{v}_{S/P} = 1195$ ft/s at 90°; $\mathbf{a}_{S/P} = 32.2$ ft/s^2 at -90°.
11.67 44.1 km/h.
11.69 (*a*) 1.970 m/s^2. (*b*) 1.502 m/s^2.
11.71 2.44 ft/s^2.
11.73 (*a*) 25.3 ft. (*b*) 11.39 ft.
11.75 (*a*) $\mathbf{v}_B = -(25 \text{ mm/s})\mathbf{e}_r + (25 \text{ mm/s})\mathbf{e}_\theta$
(*b*) $\mathbf{a}_B = (75 \text{ mm/s}^2)\mathbf{e}_r - (100 \text{ mm/s}^2)\mathbf{e}_\theta$
(*c*) $\mathbf{a}_{B/OA} = (50 \text{ mm/s}^2)\,\mathbf{e}_r$.
11.77 $(v_o/h)\sin^2\theta$.
11.79 437 mi/h at -5.9°.
11.81 Toward *A*.

CHAPTER 12

12.1 (*a*) 4.00 s. (b) 0.637.
12.3 (*a*) $\mathbf{a}_A$ = 8.4 ft/s^2 up; $\mathbf{a}_B$ = 4.2 ft/s^2 down. (b) 152.2 lb.
12.5 (*a*) 1.2 m/s right. (*b*) 0.6 m/s, left.
12.7 135.3 ft.
12.9 1.754 m/s left.
12.11 (*a*) 0.742W. (*b*) 0.940W.
12.13 (*a*) 1.207 m/s. (*b*) 21.8° and 158.2°.
12.15 3.5 m/s.
12.17 2.86 m/s.
12.19 $\mathbf{a}_A$ = 4.04 m/s^2 up; $\mathbf{a}_B$ = 0.577 m/s^2 down; $\mathbf{a}_C$ = 2.89 m/s^2 down; block C strikes first.
12.21 (*a*) F_r = -0.200 N; F_θ = 0 N. (*b*) F_r = +0.150 N; F_θ = -0.200 N.
12.23 (*a*) v_r = -2.40 m/s; v_θ = +3.60 m/s. (*b*) a_r = -54 m/s^2; a_θ = -57.6 m/s^2. (*c*) -11.52 N.
12.27 (*a*) 3750 mi/h. (*b*) 3570 mi/h.
12.29 (*a*) 770 m/s. (*b*) -75 m/s.
12.31 (*a*) $a_r = a_\theta = 0$. (*b*) 192 in./s^2. (*c*) 6 in./s.
12.33 (*a*) 0.4 m/s. (*b*) a_r = +2.4 m/s^2; a_θ = 0. (*c*) 2.8 m/s^2.
12.35 (*a*) 1.804 μN. (*b*) 190.9 s.
12.37 $F = mh^2/r^3$.
12.39 1.90x10^{27} kg.
12.41 19.74 km/s.
12.43 (*a*) 26.0x10^3 ft/s. (*b*) 173.3 ft/s.
12.45 (*a*) –413 m/s. (*b*) –156.3 m/s. (*c*) –2305 m/s.
12.47 7 h 7 min 13 s.
12.49 -123.0 m/s.
12.51 28.8 mi.

CHAPTER 13

13.1 (*a*) 10.02 ft. (*b*) 12.7 ft/s. (c) 131.9 ft-lb.
13.3 (*a*) 322 ft. (*b*) F_{AB} = 5400 lb *C*; F_{BC} = 12600 lb *C*.
13.5 (*a*) 171.6 ft-lb. (*b*) 0.198.
13.7 (*a*) 800 mm/s left. (*b*) 400 mm/s right. (*c*) 7.5 N.
13.9 (*a*) 9.59 J. (*b*) F_A = 128.0 N; F_B = 134.4 N.
13.11 464 mm.
13.15 (*a*) 77.2 km. (*b*) 80.7 km.
13.17 (*a*) 3.13 m/s. (*b*) 3.5 m/s.
13.19 2.40 MW.
13.21 (*a*) 63.3 kW. (*b*) 63.3 kW.
13.23 (*a*) 36.4 hp. (*b*) 60.1 hp.
13.25 (*a*) 440 mm; 4.80 m/s. (*b*) 440 mm 2.4 m/s.
13.27 44.0 ft/s.
13.29 (*a*) 1.981 m/s. (*b*) 221 mm.
13.31 13.57 in.
13.33 1.939 lb/in.
13.35 $3l/5$.
13.37 (*a*) 67.3x10^9 ft-lb. (*b*) 154.6x10^9 ft-lb.
13.39 (*a*) 8.75 m/s. (*b*) 309 mm.
13.41 5440 km.
13.43 (*a*) 3.40 s. (*b*) 25.5 s.
13.45 (*a*) 5.49 s. (*b*) -(92.6 ft/s)**j** - (41.0 ft/s)**k**.
13.47 (*a*) 9.76 s. (*b*) F_{AB} = 5400 lb *C*; F_{BC} = 12600 lb *C*.
13.49 (*a*) $\mathbf{v}_A = \mathbf{v}_B$ = 3.30 m/s down. (*b*) 3.64 N up.
13.51 (*a*) 38.6 ft/s right. (*b*) 12.08 ft/s right.
13.53 6830 lb.
13.55 (*a*) car *B*. (*b*) 52.9 mi/h.
13.57 (*a*) 14.40 J; 4.80 N-s. (*b*) 12.00 J, 4.00 N-s.
13.59 (*a*) $\mathbf{v}'_A$ = 1 ft/s left; $\mathbf{v}'_B$ = 15 ft/s right, (*b*) 3.35 ft-lb.
13.61 $\mathbf{v}'_A = 0.502v_0$ at 55.1°; $\mathbf{v}'_B = 0.823v_0$ at -30°.
13.63 $0.425h$.
13.65 (*a*) 0.75. (*b*) 6 in.
13.67 $\mathbf{v}'_A = 0.4v_0$ at 60°; $\mathbf{v}'_B = 0.7v_0$ left.
13.69 θ'_A = 9.70°; θ'_B = 29.4°.
13.71 (*a*) 0.80. (*b*) 0.26.
13.73 (*a*) 1.272 m. (*b*) 0.498 %.

CHAPTER 14

14.1 (*a*) 3.97 m/s. (*b*) 362 m/s.
14.3 (*a*) $\mathbf{v}'_A = \mathbf{v}'_B = 3$ mi/s right; $\mathbf{v}'_C = 0$,
(*b*) $\mathbf{v}''_A = \mathbf{v}''_B = 1$ mi/s right; $\mathbf{v}''_C = 4$ mi/h right.
14.5 (*a*) $v_x = 0.75$ m/s; $v_z = -0.4375$ m/s. (*b*) $\mathbf{H}_0 = -(44.7\ \text{kg-m}^2/\text{s})\mathbf{i}$.
14.7 $\mathbf{H}_0 = -(8.45$ ft-lb-s)$\mathbf{i}$ + (16.15 ft-lb-s)$\mathbf{j}$ + (14.16 ft-lb-s)$\mathbf{k}$.
14.9 $r_A = (240$ ft)$\mathbf{i}$ + (240 ft)$\mathbf{j}$ + (2160 ft)$\mathbf{k}$.
14.11 $x = 255$ m; $y = 7.93$ m; $z = -7.00$ m.
14.13 $\mathbf{v}_C = (1650$ ft/s)$\mathbf{i}$ - (150 ft/s)$\mathbf{j}$ + (270 ft/s)$\mathbf{k}$.
14.15 (*a*) 2.21 kg. (b) 9.35 m/s.
14.17 (*a*) 18,040 ft-lb. (b) 12,020 ft-lb.
14.19 $\mathbf{v}_B = 1.485$ m/s right; $\mathbf{v}_A = 0.990$ m/s left.
14.21 (*a*) $\mathbf{v}_A = \mathbf{v}_B = v_0/2$ right. (*b*) $\mathbf{v}_A = v_0$ right; $v_B = 0$,
(*c*) $\mathbf{v}_A = \mathbf{v}_B = \mathbf{v}' = v_0/2$ right.
14.23 (*a*) 11.09 ft/s at 210°. (*b*) 2.74 ft/s right.
14.27 (*a*) $\mathbf{v}_A = 2.56$ m/s up; $\mathbf{v}_B = 4.24$ m/s at –31.9°. (*b*) 2.34 m.
14.29 (*a*) $v_B = 4.47$ ft/s, -63.4°; $v_C = 8.00$ ft/s right. (*b*) 3 ft.
14.31 (*a*) $\mathbf{v}_A = \mathbf{v}_B = 0$, $\mathbf{v}_C = v_0$ right. (*b*) $\mathbf{v}_B = \mathbf{v}_C = v_0/2$ right; $\mathbf{v}_A = 0$.
(*c*) $\mathbf{v}_A = v_0/2$ left; $\mathbf{v}_B = \mathbf{v}_C = 2v_0/3$ right.
14.33 540 N.
14.35 $Q_1 = 62.7$ gal/min, $Q_2 = 188.0$ gal/min.
14.37 5.91 kips.
14.39 23.8 m/s.
14.41

$$v = \sqrt{gh}\,\frac{e^{\left(2\sqrt{gh}/L\right)t} - 1}{e^{\left(2\sqrt{gh}/L\right)t} + 1}.$$

14.43 (*a*) 900 kg. (*b*) 3,580 m/s.
14.45 $\mathbf{D} = 252$ lb up; $\mathbf{C}_x = 194.3$ lb left; $\mathbf{C}_y = 139.9$ lb down.
14.47 (*a*) $P = qv$.

CHAPTER 15

15.1 ω_{max} = 15.08 rad/s; α_{max} = 189.5 rad/s^2.

15.3 $\mathbf{v}_B$ = -(9 m/s)**j** + (15 m/s)**k**; $\mathbf{a}_B$ = -(1020 m/s^2)**i** - (1125 m/s^2)**j** - (625 m/s^2)**k**.

15.5 $\mathbf{v}_F$ = -(0.4 m/s)**i** - (1.4 m/s)**j** - (0.7 m/s)**k**;
$\mathbf{a}_F$ = (8.4 m/s^2)**i** + (3.3 m/s^2)**j** - (11.4 m/s^2)**k**.

15.7 (*a*) 100 mm/s^2. (*b*) 111.8 mm/s^2. (*c*) 224 mm/s^2.

15.9 (*a*) $\mathbf{v}_C$ = 67.7 ft/s left; $\mathbf{a}_C$ = 24,500 ft/s^2 down.
(*b*) $\mathbf{v}_C$ = 6.77 ft/s left; $\mathbf{a}_C$ = 245 ft/s^2 at -86.8°.

15.11 (*a*) 9.6 rad/s^2 CW. (*b*) 16.04 rev CW.

15.13 (*a*) 0.509 rad/s^2. (*b*) 20.6 s.

15.15 (*a*) 16.77 s. (*b*) ω_A = 480 rpm CCW; ω_B = 360 rpm CW.

15.17 (*a*) 1.407 rad/s CCW. (*b*) 1.198 m/s at 25°.

15.19 (*a*) 3 rad/s CW. (*b*) $\mathbf{v}_A$ = (300 mm/s)**i** + (270 mm/s)**j**.

15.21 (*a*) $\mathbf{v}_A$ = (24 in./s)**i** + (12 in./s)**j**. (*b*) y = 0; x = 3 in.

15.23 (*a*) 240 rpm CCW. (*b*) 0.

15.25 ω_B = 144 rpm CW; ω_C = 80 rpm CCW.

15.27 (*a*) ω_{BD} = 0; $\mathbf{v}_D$ = 1.885 m/s, left,
(*b*) ω_{BD} = 7.26 rad/s CCW; $\mathbf{v}_D$ = 1.088 m/s, left,
(*c*) ω_{BD} = 0; $\mathbf{v}_D$ = 1.885 m/s right.

15.29 ω_{BD} = 1.5 rad/s CW, ω_{DE} = 4.00 rad/s CCW.

15.31 ω_{BD} = 10 rad/s CCW; ω_{DE} = 7.5 rad/s CW.

15.33 Instantaneous axis is vertical line located at a distance 9.19 ft from the y axis.

15.35 (*a*) 5 rad/s CCW. (*b*) 600 mm/s right. (*c*) 450 mm.

15.37 (*a*) 6.93 rad/s CCW. (*b*) 1.039 m/s left. (*c*) 2.11 m/s at 214.7°.

15.39 (*a*) 1.5 rad/s CCW. (*b*) 18 in./s up. (*c*) 11.25 in./s at 126.9°.

15.41 (*a*) ω_{AC} = 4.8 rad/s CW. (*b*) ω_{BD} = 3 rad/s CW. (*b*) 1.5 m/at 253.7°.

15.43 (*a*) 29.3 in./s at 79.6°. (*a*) 27.4 in./s at -20.5°.

15.45 (*a*) 48.9°. (*b*) 216 in./s left.

15.47 (*a*) ω_{AB} = 1.2 rad/s CCW; ω_{DE} = 0.45 rad/s CCW. (*b*) 105.0 mm/s right.

15.49 (*a*) 10 in./s^2 left. (*b*) 10 in./s^2 right.

15.51 14.04 m/s^2 at 216.7°.

15.53 $\mathbf{a}_D$ = 7,400 ft/s^2 at -45°; $\mathbf{a}_E$ = 1,529 ft/s^2 at 45°.

15.55 32.0 m/s^2.

15.57 $\omega_{DE} = \omega_0/2$ CCW; ω_{EF} = 0; $\alpha_{EF} = 3\omega_0^2/4$ CW; α_{DE} = 0.

15.59 (*a*) 1.925 rad/s^2 CW. (*b*) 3.86 m/s^2 40.3°.

15.61 (*a*) 1.974 rad/s^2 CCW. (*b*) 10.03 m/s^2 at 250.3°.

15.63
$$\alpha = \left[\frac{v_B \sin\beta}{l}\right]^2 \frac{\sin\theta}{\cos^3\theta}.$$

15.65
$$v_p = -b\omega\sin\omega t\left[1 + \frac{b\cos\omega t}{l\sqrt{1 - \frac{b^2}{l^2}\sin^2\omega t}}\right].$$

15.67 (*a*) 1.815 rad/s CW. (*b*) 16.42 m/s at -20°.

15.69 ω_{AP} = 4.21 rad/s CCW; ω_{BE} = 1.274 rad/s CCW.

15.71 (*a*) 240 m/s^2 at 128.1°. (*b*) 351 m/s^2 at 147.5°.

15.73 (*a*) 742 mm/s at 104.0°. (*b*) 876 mm/s^2 at 170.5°.

15.75 (*a*) 24.7 in./s at 104°. (*b*) 74.2 in./s^2 at -14.0°.

15.77 317 ft/s^2 at 221.8°.

15.79 (*a*) 0.291 rad/s CCW. (*b*) 0.1206 rad/s^2 CCW.

15.81 (*a*) 3.61 rad/s CCW. (*b*) 86.6 in./s at 30°. (*c*) 563 in./s^2 at 133.9°.

15.83 (*a*) ω = (2 rad/s)**i** - (1 rad/s)**j** - (3 rad/s)**k**.
(*b*) $\mathbf{v}_D$ = (22 in./s)**i** - (16 in./s)**j** + (20 in./s)**k**.

15.85 (*a*) ω = (1 rad/s)**i** - (3 rad/s)**j** + (2 rad/s)**k**.
(*b*) $\mathbf{v}_A$ = -(105 mm/s)**i** + (165 mm/s)**j** + (300 mm/s)**k**;
$\mathbf{v}_B$ = (225 mm/s)**i** + (315 mm/s)**j** + (360 mm/s)**k**.

15.87 86.4 rad/s^2 toward the East.

15.89
(*a*) $\omega = \frac{R}{r}\omega_1\mathbf{i} + \omega_1\mathbf{j}.$

(*b*) $\alpha = \left(\frac{R\omega_1^2}{r}\right)\mathbf{k}.$

15.91 $\mathbf{v}_D$ = -(4.44 in./s)**i** + (9.52 in./s)**j** - (5.71 in./s)**k**.
$\mathbf{a}_D$ = -(3.24 in./s^2)**i** - (1.109 in./s^2)**j** + (1.331 in./s^2)**k**.

15.93
(*a*) $\alpha = -\frac{\omega_1\omega_2}{\sqrt{5}}\mathbf{k}.$

(*b*) $\mathbf{v}_C = r\omega_2\mathbf{k}.$

(*a*) $\mathbf{a}_C = 2r\left(\omega_1\omega_2 + \frac{\omega_2^2}{\sqrt{5}}\right)\mathbf{i} - r\frac{\omega_2^2}{\sqrt{5}}\mathbf{j}.$

15.95
(*a*) $\omega_{\text{spin}} = \frac{\sin\gamma}{\sin\beta}\omega_1.$

(*b*) $\omega = \frac{\sin(\gamma-\beta)}{\sin\beta}\omega_1.$

(*c*) $\alpha = \frac{\sin(\gamma-\beta)}{\sin\beta}\sin\gamma\,\omega_1^2\mathbf{k}.$

15.97 $\mathbf{v}_C$ = -(23 mm/s)**i**.

15.99 (*a*) $\mathbf{v}_D$ = (42 in./s)**i** - (28 in./s)**j** + (96 in./s)**k**.
(*b*) $\mathbf{a}_C$ = (336 in./s^2)**i** - (252 in./s^2)**j**.

15.101 (*a*) $\mathbf{v}_D$ = (22 in./s)**i** - (60 in./s)**j** + (25 in./s)**k**.
(*b*) $\mathbf{a}_D$ = (145 in./s^2)**i** + (192 in./s^2)**j** - (168 in./s^2)**k**.

15.103 (*a*) $\mathbf{v}_D$ = (0.6 m/s)**i** + (1.039 m/s)**j** - (1.247 m/s)**k**.
(*b*) $\mathbf{a}_D$ = -(16.9 m/s^2)**i** + (4 m/s^2)**j** - (9.6 m/s^2)**k**.

15.105 (*a*) α = -(0.27 rad/s^2)**i**.
(*b*) $\mathbf{v}_D$ = (6.24 in./s)**i** - (3.6 in./s)**j** - (16.8 in./s)**k**.
(*c*) $\mathbf{a}_D$ = -(11.70 in./s^2)**i** - (2.81 in./s^2)**j** - (7.48 in./s^2)**k**.

15.107 (*a*) $\omega = \omega_1\mathbf{i} + \omega_2\mathbf{k}$.

$\alpha = -\omega_1\omega_2\mathbf{j}$.

(*b*) $\mathbf{v}_D = r\,\omega_2\mathbf{j} - r\,\omega_1\mathbf{k}$.

$\mathbf{a}_D = r\omega_2^2\mathbf{i} + r\omega_1^2\mathbf{j} + 2r\,\omega_1\omega_2\mathbf{k}$.

15.109 (*a*) $\alpha = -(600\text{x}10^{-6}\text{ rad/s}^2)\mathbf{k}$.

(*b*) $\mathbf{v}_D = -(112.6\text{ mm/s})\mathbf{i} + (168.9\text{ mm/s})\mathbf{j} + (172.5\text{ mm/s})\mathbf{k}$.

(*c*) $\mathbf{a}_D = (6.90\text{ mm/s}^2)\mathbf{i} - (5.18\text{ mm/s}^2)\mathbf{j} + (7.32\text{ mm/s}^2)\mathbf{k}$.

15.111 (*a*) $\alpha = -(0.262\text{ rad/s}^2)\mathbf{k}$.

(*b*) $\mathbf{v}_B = (313\text{ ft/s})\mathbf{k}$.

$\mathbf{a}_B = (62.7\text{ ft/s}^2)\mathbf{i} - (822\text{ ft/s}^2)\mathbf{j}$.

15.113 (*a*) $\mathbf{v}_A = (0.81\text{ m/s})\mathbf{i} - (0.72\text{ m/s})\mathbf{j} + (0.76\text{ m/s})\mathbf{k}$.

(*b*) $\mathbf{a}_A = (0.64\text{ m/s}^2)\mathbf{i} + (1.392\text{ m/s}^2)\mathbf{j} - (1.920\text{ m/s}^2)\mathbf{k}$.

CHAPTER 16

16.1 (*a*) 8.05 ft/s^2. (*b*) **B** = 10.27 lb up; **C** = 5.73 lb up.
16.3 (*a*) 3.33 m/s^2 right. (*b*) **A** = 2.16 N up; **B** = 12.55 N up.
16.5 (*a*) 2.86 m/s^2 right. (*b*) 30.7 N.
16.7 (*a*) 0.349g, (*b*) 4.
16.9 5.48 m/s^2, right.
16.11 F_{AD} = 18.84 N T; F_{BE} = 7.65 N T.
16.15 54.0 lb.
16.17 (*a*) 16.1 ft/s^2 at -30°. (*b*) $\mathbf{a}_B$ = 20.1 ft/s^2 down; $\mathbf{a}_P$ = 40.2 ft/s^2 at -30°.
16.19 57.9 N-m.
16.21 (*a*) 62.8 rad/s^2 CCW. (*b*) 57.1 rad/s^2 CCW.
16.23 0.904 r.
16.25 100.8 lb.
16.27 (*a*) 6.37rad/s^2 CW. (*b*) 1.912 m/s up.
16.29

$$I_R = \left(n + \frac{1}{n}\right)^2 I_0 + n^4 I_C.$$

16.31 (*a*) α_B = 49.1 rad/s^2 CCW; α_A = 11.77 rad/s^2 CCW.
(*b*) **C** = 41.2 N up; $\mathbf{M}_C$ = 1.130 N-m CCW.
16.33

$$\alpha = \frac{2\mu_k g \sin\phi}{r(\sin 2\phi - \mu_k \cos 2\phi)} \quad \text{CCW.}$$

16.35 (*a*) 16 rad/s^2 CCW. (*b*) 2.4 m/s^2 right. (*b*) 300 mm from B.
16.37 $\mathbf{a}_A$ = 0.430 m/s^2 up; $\mathbf{a}_B$ = 1.966 m/s^2 up.
16.39

(*a*) W. (*b*) $\alpha = \dfrac{rg}{\bar{k}^2}$ CCW.

16.41 (*a*) 702 lb. (*b*) 0.377 rad/s^2 CCW.
16.43 (*a*) 5.196g/L CCW. (*b*) 3.50g at 188.2°. (*c*) 1.803g at –16.1°.
16.45 (*a*) 80 mm. (*b*) 41.7 rad/s^2 CW.
16.47 (*a*) 244 kN. (*b*) 187.0 kN.
16.49 41.9 N-mm CCW.
16.51 (*a*) 1.061g/b CW. (*b*) 3g/2 down. (*c*) W/4 up.
16.53 (*a*) 29.9 rad/s^2 CW. (*b*) $\mathbf{A}_x$= 28.0 lb left; $\mathbf{A}_y$= 4.29 lb up.
16.55 (*a*) 53.4 N down. (*b*) $\mathbf{A}_x$= 54.0 N left; $\mathbf{A}_y$= 111.5 N up.
16.57 (*a*) 43.3 lb-ft CCW. (*b*) 3.31 lb at 30°.
16.59 1.111 m/s^2 left.
16.61 (*a*) 75.5 rad/s^2 CW. (*b*) 7.55 m/s^2 down.
16.63 $\mathbf{a}_A$= 4.19 ft/s^2 left; $\mathbf{a}_B$= 3.26 ft/s^2 left.
16.65 (*a*) 13.07 rad/s^2 CW. (*b*) 9.06 N up.
16.67 (*a*) 7.81 lb left. (*b*) 5.71 lb.
16.69 (*a*) 2.98 rad/s^2 CW. (*b*) 1.396 lb at 130°.

CHAPTER 17

17.1 1.189r.

17.3 40.6 rev.

17.5 (*a*) 11.39 rev. (*b*) –(5.26 lb)**j**.

17.7 1.079 m/s down.

17.9 (*a*) 53.4°. (*b*) $0.767\sqrt{gr}$ at –53.4°.

17.11 689 mm.

17.15 (*a*) 7.43 ft/s down. (*b*) 4.0 lb.

17.17 (*a*) 421 mm/s left. (*b*) Each reaction 141.5 N up.

17.19 3.60 m/s right.

17.21 3.23 ft/s left.

17.23 (*a*) 39.8 N-m. (*b*) 99.5 N-m. (*c*) 249 N-m.

17.25 4.11 rad/s CW.

17.27 (*a*) 2.70 m/s down. (*b*) 24.2 N.

17.29 (*a*) $gt/2$ down. (*b*) $2gt/3$ down.

17.31 11.71 ft/s up.

17.33 (*a*) Pipe rolls without sliding on plate.
(*b*) $\mathbf{v} = 0.849$ m/s right; $\omega = 7.08$ rad/s CCW; $\mathbf{v}_{Plate} = 1.698$ m/s right.

17.35 2.78 m/s left.

17.37 $2\omega_0/5$

17.39 (*a*) 4.98 ft/s right. (*b*) 1.246 ft/s left.

17.41 (*a*) 12 in. (*b*) 6 ft/s right.

17.43 242 mm/s right.

17.45 $\omega_2 = \dfrac{6}{7}\dfrac{\mathbf{v}_1}{L}$ CCW; $\mathbf{v}_2 = \dfrac{3\sqrt{2}}{7}\mathbf{v}_1$ at 225°.

17.47 $\omega_1/2$

17.49 $L/\sqrt{3}$

17.51 (*a*) 1.918 N-s. (*b*) 2.35 N-s.

17.53 (*a*) $0.7v_1$ right. (*b*) $1.8\ v_1/L$ CCW. (*c*) $0.3v_1$ right.

17.55 3.01 ft/s at 210°.

17.57 $\omega_{AB} = 2.65$ rad/s CW; $\omega_{BC} = 13.25$ rad/s CCW.

CHAPTER 18

18.1 $\frac{1}{12}ma^2\omega\ (3\mathbf{j} + 2\mathbf{k})$.

18.3 (*a*) – (1.041ft · lb · s)**i** + (1.041ft · lb · s)**j** + (2.31 ft · lb · s)**k**.
(*b*) 147.5^o.

18.5 (*a*) $(3F\Delta t/2ma)(\mathbf{i} - 8\mathbf{j})$. (*b*) Axis through A,
in *xy* plane, forming 277.1^o with x axis.

18.7 42.9^o.

18.9 -(0.625 rad/s)**i** - (0.281 rad/s)**j** + (0.0844 rad/s)**k**.

18.11 (*a*) 22.1 in. (*b*) ω_y= 1.950 rad/s, ω_z= 0.406 rad/s.

18.13 29.1 ft · lb.

18.15 - 161.4 ft · lb.

18.17 $\frac{1}{6}ma^2\omega^2\ \mathbf{i}$.

18.19 $-(\frac{1}{6}mb^2\omega^2 \sin 2\beta)\mathbf{k}$.

18.21 **A** = -(13.08 lb)**i** -(13.08 lb)**j** ; **B** = (13.08 lb)**i** -(13.08 lb)**j**.

18.23 (*a*) (10.56 mN□m)**k**.
(*b*) **A** = -**B** = (4 mN)**i** + (8 mN)**j**.

18.25 (*a*) (6 rad/s)**k**.
(*b*) **A** = -**B** = -(20 mN)**i** + (20 mN)**j**.

18.27 5.18 lb up.

18.29 (*a*) $\cos^{-1}(3g/2l\omega^2)$. (*b*) $0 \le \omega \le \sqrt{3g/2l}$.

18.31 $\sqrt{8g/11a}$.

18.33 $\mathbf{A} = -\frac{1}{4}ma\omega_1\omega_2\,\mathbf{k}$; $\mathbf{B} = \frac{1}{4}ma\omega_1\omega_2\,\mathbf{k}$.

18.35 (*a*) (7.97 N□m)**j**.
(*b*) **A** = -(9.00 N)**i** - (25.5 N)**k**,
M_A = (3.75 N□m)**i**.

18.37 (*a*) $\frac{1}{4}ma^2\alpha_1\,\mathbf{i}$.
(*b*) $\mathbf{A} = -\frac{1}{4}ma^2\omega_1\omega_2\,\mathbf{k}$; $\mathbf{B} = \frac{1}{4}ma^2\omega_1\omega_2\,\mathbf{k}$.

18.39 (*a*) 30°. (*b*) 25.3^o. (*c*) 36.5^o.

18.41 (*a*) 128.3 rad/s. (*b*) 2.17 in.

18.43 Precession axis: $\theta_x = 39.9^o$, $\theta_y = 127.9^o$,
$\theta_z = 79.4^o$; precession, 0.458 rad/s;
spin, 0.273 rad/s.

18.45 Precession axis: $\theta_x = 90^o$, $\theta_y = 26.0^o$,
$\theta_z = 64.0^o$; precession, 0.845 rad/s
(retrograde); spin, 0.1591 rad/s.

18.47 (*a*) $\sqrt{17g/11a}$. (*b*) $\sqrt{44g/17a}$.

18.49 (*a*) $\sqrt{15g/11a}$.
(*b*) $\dot{\phi} = 2\sqrt{20g/33a}$; $\dot{\psi} = \sqrt{20g/33a}$.

18.51 (*a*) (0.825 kg·m²/s)**i** + (0.592 kg·m²/s)**j**. (*b*) 1.510 J.

18.53 14.01 rev/h.

CHAPTER 19

19.1 7.5 mm, 3.18 Hz.

19.3 5.08 ft/s, 77.3 ft/s^2.

19.5 11.04 mm or 0.435 in.

19.7 (*a*) 0.0540 s. (*b*) v = 5.18 ft/s up; a = 26.8 ft/s^2 up.

19.9 (*a*) 3.86 lb. (*b*) 52.5 lb/ft.

19.11 3.41 kg.

19.13 (*a*) $\ddot{x} + (k/2ml^2)x^3 = 0$, not SHM.
(*b*) $\ddot{x} + (T_0/ml)x = 0$, SHM.

19.15 (*a*) 3.57 kg. (*b*) 38.1 kg.

19.17 (*a*) 0.349 s. (*b*) 0.450 m/s.

19.19 (*a*) 0.450 s. (*b*) 21.0 in./s.

19.21 (*a*) 1.125 Hz. (*b*) $T_B = 76.7$ N; $T_C = 54.2$ N.

19.23 (*a*) 0.428 s. (*b*) 45.4 mm.

19.25 $2\pi\sqrt{2m/3k}$.

19.27 (*a*) and (*b*) 0.933 s.

19.29 (*a*) $2\pi\sqrt{5r/4g}$. (*b*) r/4.

19.31 (*a*) $2\pi\sqrt{h/2g}$. (*b*) $2\pi\sqrt{h/g}$.

19.33 75.5°.

19.35 (*a*) $(1/2\pi)\sqrt{2c/3g}$. (*b*) $(1/2\pi)\sqrt{4c/5g}$.

19.37 5.28 Hz.

19.39 1.700 s.

19.41 2.10 Hz.

19.43 $0.1899\sqrt{g/\ell}$.

19.45 $2\pi\sqrt{2hL/3bg}$.

19.47 $2\pi\sqrt{(\pi-2)r/g}$.

19.49 (*a*) 100 mm. (*b*) 35 mm (out of phase).

19.51 ω greater than $\sqrt{2k/m}$.

19.53 ω between $\sqrt{2g/3\ell}$ and $\sqrt{4g/3\ell}$.

19.55 283 rpm.

19.57 (*a*) 175.0 μm, in phase. (*b*) 769 μm, out of phase.
(*c*) infinity (resonance).

19.59 1.714 mm (out of phase), 2.40 mm (out of phase).

19.61 0.343 in. (out of phase), 2.4 in. (in phase).

19.63 0.587 Hz, 0.455 Hz.

19.65 (*a*) 1319 lb·s/ft. (*b*) 0.1107 s.

19.67 $\sqrt{1-2(c/c_c)^2}$.

19.69 (*a*) 0.0481 in. (out of phase). (*b*) 0.0234 in.

19.71 (*a*) 0.0788. (*b*) 319 N·s/m.

19.73 1.991 kN·s/m.

19.75 (*a*) 297 rpm. (*b*) 223 rpm.
(*c*) 0.268 in. for part *a* and 0.303 in. for part *b*.